From Energy Dreams to Nuclear Nightmares

From Energy Dreams to Nuclear Nightmares

Lessons from the anti-nuclear power movement in the 1970s

Horace Herring

JON CARPENTER

Acknowledgement

Extracts from *Nuclear Fear: A History of Images* by Spencer R. Weart are reprinted by permission from the publisher, Cambridge, Mass.: Harvard University Press, Copyright © 1988 by the President and Fellows of Harvard College.

Our books may be ordered from bookshops or (post free) from Jon Carpenter Publishing, Alder House, Market Street, Charlbury, England OX7 3PH

Credit card orders should be phoned or faxed to 01689 870437 or 01608 811969

First published in 2005 by
Jon Carpenter Publishing
Alder House, Market Street, Charlbury, Oxfordshire OX7 3PH
☎ 01608 811969

© Horace Herring 2005

ISBN 1 897766 99 8

Printed in England by Antony Rowe Lrd., Eastbourne

Table of Contents

Acknowledgements

To Phil for getting me started
and to Dave for keeping me going.
To the library for filling my head and
to 'Janet' for providing me with the space to think:
from the empty blankness of the M6 junctions 1 to 4
to the mellow hills of rural Shropshire
with only *Echoes* to relieve the hours of darkness.
Above all to Olive for her chiding kindness.

Preface

Conflicting utopias

Mankind has always dreamed of futures in which life was improved via, for example, the adoption of more progressive political arrangements, or more latterly, radical new technologies. Certainly, civil nuclear power was initially seen by many in utopian terms, as the harbinger of social and economic progress. The fact that it had started life as a weapons technology only strengthened the commitment of progressives to showing how technology could improve man's lot – 'atoms for peace' was the cry. However, this view was opposed by a range of groups in society who were worried about the impacts of nuclear power and who also had different views of the future. For example, there were those who hankered after a more traditional society – some of whom seemed to share a quasi-romantic dream of a return to a pre-industrial arcadia.

Linked to this group to varying degrees, there were those whose main concern was the protection and conservation of the natural environment, and those who were more concerned about the whole project of capitalistic social and economic change. But the opposition to nuclear was not just a reaction to change – there was also a more progressive element, represented by those who argued for alternative more environmentally appropriate technologies, which they saw as a much better bet. They had a vision of a future based on solar energy in its various forms – what we now call a sustainable energy future, based on the use of renewable energy sources. This vision is one that was shared by most of the new more radical environmental groups (e.g. Friends of the Earth and then Greenpeace) that emerged in the 1970's and who took the lead in opposing nuclear power.

In this book, Horace Herring looks at how the various rival utopias were established and how they developed, interacted and changed. As he describes, initially, pro-nuclear utopians were ranged against reaction from anti-nuclear 'conservatives' – not just environmental preservationists, but also these who wanted to protect specific industrial interests, notably the coal lobby. Subsequently, the nuclear progressives also found themselves challenged by the pro-renewables lobby, parts of which also had a strong commitment to an alternative more decentralised society, a view also underpinning much of the burgeoning green movement and shared by many of the radical environmental groups. Although as Horace describes, different elements have dominated in the opposition in each period, one way or another, for most of the last few decades, the nuclear dream has thus been kept in check.

However, what is interesting in developments in recent years is that the overlap within the anti-nuclear/pro-renewables consensus, between the 'conservationist' perspective and the progressive renewable energy and green activists perspectives, may now be reducing. As the reality of renewable energy technology becomes clearer, some conservationists are beginning to object to it, particularly in the case of wind farms, ostensibly on environmental grounds. Indeed, some have even begun to rethink their opposition to nuclear power – in the belief that renewables will not be viable or acceptable on a scale sufficient to respond to the threat climate change. For its part, the renewables lobby, backed by the progressive environmental groups, argues against what it sees as an unnecessary and dangerous retreat – instead they insist that renewables coupled with energy conservation can lay the basis for a sustainable future. But clearly we are faced with another twist in the ongoing battle over which vision of the future to steer towards – a battle nowadays sharpened by the urgent need to respond to climate change.

Dave Elliott
Professor of Technology Policy
The Open University

Chapter 1

Introduction

The year 2005 is an ironic time to examine the myths of the nuclear age. Just over fifty years ago (on 8 December 1953) President Eisenhower gave his Atoms for Peace speech, which launched the nuclear age, heralding it as the 'dream of the future'. However fifty years later instead of atomic success we see the financial collapse of British Energy, the UK's main nuclear generator. Underlining this fall from grace is the stated refusal of the current Labour government, such as in its latest White Paper on Energy, to support the construction of any more nuclear power stations. Thus we are now witnessing the end of one of the most famous myths of the nuclear age that has lasted for fifty years, that electricity from nuclear power stations would become 'too cheap to meter'.

Ever since the dawn of the atomic age, there has been scattered opposition to the use of nuclear power. This opposition was in the form both of intellectuals writing articles or speeches and local communities and conservation bodies opposing the building of nuclear power stations in rural areas. However these protests were ignored and marginalized: there was no organization or movement willing to embrace such anti-nuclear ideas.

Thus those opposing nuclear power had to wait until a social movement existed that was willing to listen to their arguments, and such movements did not exist until the student and alternative movements in the 1960s. Again there had to be a bridge between the old anti-nuclear criticism and these new movements: this was provided by such established writers as Lewis Mumford, Barry Commoner and Fritz Schumacher. But more important were a new generation of anti-nuclear writers and activists, particularly in the USA, people who were located within both existing radical student movements and the emerging anti-nuclear power movement. They had both the ideas and the constituency of supporters. They were thus in a position to build a large-scale movement, and their ideas and methods spread to the UK.

FoE account

The main account of the anti-nuclear power campaign in Britain, in the mid to late 1970s, has been written by Walt Patterson, who led the FoE campaign, and who published accounts of it in such books as his *Going Critical: an Unofficial History*. His account has been uncritically accepted by the FoE biographer Robert Lamb in his book *Promising the Earth*. In these accounts, the story is cast as David versus Goliath, of a small band of dedicated activists based in the FoE London office taking on the might of the nuclear state. There is scant mention of any other groups involved, except that they got in the way and 'muddied the waters'. This was a battle of intellects, of rationality and scientific debate fought in newspaper columns, before parliamentary committees and at the final and epic battle of the Windscale Inquiry in 1977. This, it was believed, was lost due to the incompetent and emotional arguments put forward by the rabble of environmentalists who tagged along behind FoE.

There is furthermore the view that the campaign against nuclear power in Britain only started with FoE's campaign in the mid-1970s. Walt Patterson commented in a 1979 paper on the history of 'environmental' involvement in nuclear policy that it was 'important to note that until late 1973 there was no coherent national public "campaign" challenging any British nuclear plan... in Britain adverse comment had been sporadic and unspecific: one-off media coverage of foreign issues, occasional letters to the editor, and passing references in various manifestos and reports'. Roger Williams, the nuclear historian, writing contemporaneously agreed and remarked that 'the British nuclear opposition first effectively mobilised in respect of the 1974 thermal reactor decision – occasional earlier developments had wholly failed to stimulate public interest. Friends of the Earth were from the outset in the van of this opposition'.

These early accounts may be good on policy and technical issues, but are I believe poor environmental history, as they ignore the continuity of anti-nuclear protest dating back many decades and the efforts of a diverse network of activists who took up anti-nuclear campaigning before FoE did. While it may be true that there was no public campaign, in the sense of one run by a national campaigning organization such as FoE, there had been since the first public inquiry in 1956 a sustained criticism of nuclear power, and sometimes vigorous local opposition. Perhaps most importantly there had never been complete public confidence in atomic energy, despite the intense propaganda in its favour. There was always an undercurrent of suspicion and pessimism centred on emotional fears over the 'atomic' future. This manifested itself in the popularity of 1950s science fiction literature describing post-nuclear war societies, often as a return to barbarism but sometimes a return to the pastoral ideal. It was also evident in the rise of the anti-bomb movement in the late 1950s which was to serve as a model of direct action and a source of activists for the future environment movement in the late 1960s. Pessimism here was constructive; it brought about a desire for change and generated alternative visions of the future.

Continuity

There is continuity in all social movements, caused by the existence of a network of activists, both inter-generational and activity-based. This concept of networks is used by both Derek Wall and Philip Conford in their books. Wall, a green historian and sociologist, in his history of the radical group Earth First! links its rise in the 1990s to the existence of an anti-roads movement in the 1970s. Conford traces the rise of the organic movement post-war to a network of Christian thinkers laying the intellectual foundations in the 1920s and 1930s.

In this book I will argue that new groups, such as FoE in 1970, did not just spring into life, one day people didn't wake up and say we're now 'ecologists'. Groups and movements have roots in older and experienced activists who often split from existing groups to found new ones. For instance David Brower split from the Sierra Club and started FoE in the USA in 1969, and in 1970 toured Europe to establish branches. In Paris Brower asked the help of Edwin Mathews, who by chance met Barclay Inglis, who recruited student activists to start up the UK branch office. New groups then recruit from existing political and social networks, which may be looking for a new cause to support. The environmental movement drew heavily from the existing peace and ex-CND activists, from student organizations such as the Young Liberals, and from the 'underground' or so-called 'alternative society'. It also drew upon activists from existing environmental groups such as the Conservation Society or Conservation Corps. New groups obviously attract young supporters but they are often run by older activists with previous experience of campaigns.

And there had been many conservation campaigns in the 1960s, all the way from the battle over the Cow Green reservoir in the early 1960s to the campaign against RTZ mining in Snowdonia (which was the first campaign for Amory Lovins and Walt Patterson). New groups not only draw on the resources of experienced activists but also draw on 'campaign craft' or what Derek Wall calls 'repertoires of action' or 'sets of tactics'. These are often drawn from outside the environmental movement, and give new groups a distinctive image when they are adopted in an unfamiliar context. Three new tactics FoE adopted were direct action, the street party or festival, and use of alternative media.

The roots of direct action lay in the peace and student movement, whilst street parties and alternative media (or 'underground press') were elements of the 1960s counterculture. The genius of FoE was to evolve a campaigning style that drew on the best of these three movements: the non-violence and ethics of the peace movement, the intellectual rigour of the student movement, and the friendly anarchism of the counterculture.

Birth of anti-nuclear activism in the UK

In the UK in late 1974 a few activists from within the Conservation Society (ConSoc) were the first people to start an anti-nuclear power campaign, with an anti-nuclear power petition. They received little support from within their own organization or from outside organizations such as FoE but attracted nearly 9,000 signatories. Their attempt on 22 March 1975 at a demonstration in London, when they presented their petition to Downing Street, attracted only 90 people, with few people from FoE attending. FoE at that time was campaigning and demonstrating on 'Save the Whale', packaging and bottle recycling, and cycling. A few of the FoE staff, such as Walt Patterson and Amory Lovins, were involved on the nuclear power issue, but mainly at the technical level over reactor choice and at public inquiries. Patterson had for some time been trying to interest local FoE groups in nuclear issues, but without much success.

The public and existing movements – the new left, the counterculture, feminists etc. – were not interested or concerned about nuclear issues. If anything they still all shared the common cultural position of support for nuclear power as a desirable feature of technological modernization. Perhaps the first group to adopt and promote the anti-nuclear position was the alternative technology (AT) movement, based around the magazine *Undercurrents*. Their issue no. 9 (in early 1975) had a 16-page section on nuclear power. From then on they carried an increasing number of stories on anti-nuclear issues, as did the rest of the environmental press (*The Ecologist* etc.) and new left/radical science publications (*New Statesman*, *Science for People* etc.).

The influence of the *Undercurrents* editorial collective was considerable as they were both activists and intellectuals, able to shape and report the anti-nuclear debate, and they had close links with existing social movements. They were in the classic sense 'movement intellectuals' being both shaped by and shaping the emerging movement: their ideas were formed from within it (see Eyerman and Jamison 1990). But for them to create a mass movement there had to be two essential factors: a change in the cultural climate towards nuclear power, and an issue about which they could campaign. In the mid-1970s the UK anti-nuclear and the environmental movement became increasingly new left politically with the influx of old student activists, replacing the old left/liberal values of the Conservation Society. The emphasis became on street action, not committee action: on demonstrations not submissions. Younger members of ConSoc drifted away to more radical and active groups such as FoE, and then to local anti-nuclear groups such as Half Life, SCRAM, etc. Anti-nuke was where the action was. Finally the public at large had lost their enthusiasm for the nuclear dream. Nuclear power in the form of the AGR had been a big failure, and the nuclear establishment was split over the choice of a new reactor system. The mass media, previously fawning before nuclear, was now ready to carry critical, and even anti-nuclear, articles.

Critical events

If they are to mobilize their members, all movements need a critical spark, some event or outrage about which they can coalesce. For the UK anti-nuclear movement this was the dramatic front-page story in the *Daily Mirror* in October 1975 about BNFL's plan to build a reprocessing plant at Windscale in Cumbria. Up till then it had been all talk, at House of Commons committee meetings, at public inquiries, and at nuclear debates. Unlike other countries in Europe there was little of concrete substance to protest at: no nuclear stations were under construction. The Windscale announcement offered campaigners the chance to protest at something real, at something that was going to be built. Furthermore reprocessing by producing plutonium offered campaigners the chance to widen the debate to include issues of nuclear proliferation, illicit building of atomic bombs, and long-term waste disposal.

FoE was ready for this challenge. It had not only grown in numbers to become the biggest and best known environmental group, but its anti-nuclear campaigners had established their technical credentials amongst the mass media and built up links to the new left and the environmental movements. It was now in a position to launch a mass anti-nuclear campaign. Its first (public) action was in April 1976 to hire a train to carry four hundred supporters up to Cumbria and to demonstrate with other groups outside the Windscale plant. About a thousand people attended. FoE, increasingly joined by other groups, continued their campaign, calling for a public inquiry, which was held from June to October 1977. The decision to go ahead with the plant in March 1978 provoked massive indignation and the largest anti-nuclear demonstration to date of some twelve thousand on 29 April 1978.

The Windscale campaign, although a 'failure' for the environmental movement, did however educate a whole new generation of activists in 'repertoires of action', which were to be used in the anti-nuclear campaign in the 1980s, which in turn provided a new generation for the anti-roads campaign of the 1990s. It also demonstrated how:

- A small publicity-conscious group with good media links, such as FoE, can exploit opportunities and become the dominant group, albeit at some cost to their credibility with activists.
- In building the anti-nuclear campaign FoE had to link up with existing networks but these damaged its public image for respectability.
- FoE's technical competence did not win the nuclear argument or public inquiries, but did attract media praise and build its confidence. Technical incompetence by the Windscale Appeal, on the other hand, brought ridicule and internal recrimination, and hastened the decline of ConSoc.

The 1970s anti-nuclear power campaigners had little success in influencing short-term nuclear policy. Their main achievement however was long-term, in making anti-nuclear feeling an integral part of modern environmentalism, which has generally been adopted by most European governments in the form of their support for renewable rather than nuclear energy. To change that requires redefining environmentalism towards the old preservation ideas (or perhaps the deep green approach) of the protection of 'nature' rather than 'man', together with the fading of old images of nuclear apocalypse.

Heed activists not theory

This book is influenced by my criticism of academics. In search of explanations for the rise of the environmental movement, I concentrate not on intellectual texts but on activist literature. For in my 1996 review of David Pepper's book *Modern Environmentalism* I complained:

> What is the relation between the -isms and the -ists? That is, how influential and widespread are the ideas, the environmentalism, amongst the people who do the action, the environmentalists? How many dreadlocked protesters against the Newbury by-pass have read Bookchin, how many little old ladies demonstrating against animal exports subscribe to *The Ecologist*?
>
> What is missing from this book is the voice of the activists, and their little magazines and pamphlets that rarely appear in university libraries. They are the voice of modern environmentalism, not the books by Edward Goldsmith or Murray Bookchin. Their voice has not been heard as their history of environmental activism since the 1960s has not yet been written.

Since then several works have appeared along these lines and my inspiration here are the books *Senseless Acts of Beauty* and *Earth First!* by respectively the activists and academics George McKay and Derek Wall. These books draw on a wide range of sources: activist papers and diaries, interviews and even record lyrics. This book follows that historical tradition in giving primacy to the voice of activists rather than the few intellectuals who claimed to speak in their name.

Environmental history

The roots and shades of the environmental movement are deep and complex and are beyond the scope of this book. I feel we must develop an historical perspective on how we have arrived at our present state and it is this that the discipline of environmental history, to which this book subscribes, seeks to do.

Environmental history is an academic discipline that focuses on the history of the environment and the history of environmentalism. One of its founding fathers, Donald Worster, has defined it in his book *The Wealth of Nature* (1993) as 'the interdisciplinary study of the relations of culture, technology and nature through time'. It stresses the importance of continuity of events, and the linkages between eras. As Samuel Hays commented in his recent book *A History of Environmental Politics Since 1945*:

> Events are embedded not only in context; they are also embedded in the flow of time. Each succeeding event or circumstance is related closely to a preceding one. It is difficult, if not impossible, to understand a point in time without placing it in its evolutionary context. This is not just a view of history that links events past to events present. More important, it is an understanding that emphasizes the way one thing 'grows out of' another, it arises from the old and then becomes something a bit different.

The second prong of environmental history, the history of 'environmentalism' or of 'environmental movements', is not so well developed. Generally the histories written of environmental groups have either been by activists or their supporters, such as Robert Lamb's *Promising the Earth* on FoE. There are few critical and original works. Those that are, such as Anna Bramwell's *Ecology in the 20th Century*, have proved controversial, while others like Joe Weston's *The FOE Experience*, Meredith Veldman's *Fantasy, the Bomb and the Greening of Britain*, and Martin Holdgate's *The Green Web* attract little attention.

However 'green history' is popular when the 'lessons from history' (on soil erosion, deforestation, pollution etc.) are used to support green activism, such as in the bestseller *Green History* (1991) by Clive Ponting. Another more academic approach to this genre is the writings of Derek Wall, a green activist and prolific green historian, whose approach to history is to ask movement activists how they got involved, to read movement literature, and to reassess his own experience as an activist. Similarly Joe Weston, an FoE activist in the 1970s and 1980s, writing in *The FOE Experience* believes that 'in attempting to locate groups within a theoretical knot-hole it is so often forgotten that those groups are made up of individuals and that some individuals can have a massive influence over the character of a whole organization'.

He believes that the examination of politics and political organizations from the 'inside' makes possible a far greater insight and understanding of internal clashes, faction fighting and consequent organization change than any empirically based 'scientific' study could ever achieve. Weston in his study of FoE argued that it is only through 'the witnessing and reporting of these clashes and changes that a true "taste" of Friends of the Earth can be transmitted and understood and as a result a fuller understanding of green politics as a whole can be achieved'.

I aim to follow the tradition of environmental history by writing in clear prose that is accessible to the non-specialist reader. My model here for clarity and readability is a recent book *The Origins of the Organic Movement* by Philip Conford; although one reviewer complained it had a 'propensity to under-theorise'. This book will plead guilty to that academic crime as well as the propensity to have footnotes which are intended as a help for future researchers. Those who wish to see more and fuller references should consult my thesis.[1] I write in the knowledge that there is little written on British post-war environmental activism in the 1960s and 1970s and in the belief that there should be much more.

This book is mainly about explaining the reasons for the growth of the anti-civil nuclear power movement in the UK from 1955 to 1979. It challenges some of the existing histories and associated explanations of the processes involved, and provides new explanations as to how and why, and indeed, when the anti-nuclear movement emerged. The period chosen covers the time from local opposition at the first public inquiry into a proposed nuclear reactor at Bradwell in 1956, to the impact of the Windscale Inquiry report on the national anti-nuclear power movement in 1978. A cut-off date of 1979 is selected as this marks the divide between the moderate tactics adopted by the then largest environmental groups, Friends of the Earth and the Conservation Society, and the more militant direct action approach of new organizations, such as Greenpeace, the Torness Alliance and the Anti-Nuclear Campaign. The story focuses on the battle in the mid-1970s over THORP (Thermal Oxide Reprocessing Plant) at Windscale in Cumbria, a plant designed to reprocess spent nuclear fuels and obtain plutonium to fuel the next generation of nuclear reactors planned for the 1980s – the Fast Breeder Reactor (FBR). This book is not a policy history of which there are many, but a history of activism based on accounts collected from the contemporary literature and from the activists.

I aim to give a generally new account of the anti-nuclear power movement and show that anti-nuclear feeling arose from:

- Long-standing public ambivalence over atomic energy.
- Disillusionment with claims of nuclear utopia.
- The continuity of protest over the building of nuclear reactors.
- A small network of activists, from a wide variety of backgrounds.

I believe that an anti-nuclear power movement emerged in the UK in the 1970s because of long-standing concerns of a minority over the dangers of atomic energy and the continuity of opposition to the building of nuclear power stations. The divisions within the nuclear establishment over reactor choice in the UK (and radiation and safety standards in the USA) gave new radical groups – the environmentalists – an opportunity to build a movement using new tactics of protest based on the counterculture of the 1960s.

It seeks to answer the question: Why an anti-nuclear power movement instead of an anti-coal or anti-asbestos movement? What was it about nuclear power that generated such opposition? Its environmental impact, its cost, its prospects or its symbolism? Finally it asks whether wind power in the twenty-first century could fall victim to the same forces that opposed the expansion of nuclear power thirty years ago.

Chapter 2

The saga of reprocessing

The British nuclear industry was founded to produce plutonium for nuclear weapons. Since its creation plutonium has lain at the centre of its activities. Windscale (renamed Sellafield in the 1980s) in west Cumbria was the main site at which plutonium production reactors and reprocessing plants were constructed for Britain's military programme. This was discreetly acknowledged by contemporary accounts. For instance Kenneth Jay, in his book *Calder Hall* published by the UKAEA to commemorate the commissioning of Britain's first nuclear reactor in October 1956, wrote:

> Major plants built for military purposes such as Calder Hall are being used as prototypes for civil plants. The plant has been designed as a dual-purpose plant to produce plutonium for military purposes as well as electric power.

By the mid-1960s there were sufficient stocks of plutonium for nuclear weapons but plutonium continued to be produced in the belief that large quantities would be required for the Fast Breeder Reactor (FBR), which was the ultimate goal of nuclear scientists. As Alvin Weinberg, the most articulate proponent of nuclear power, remarked: 'Our main belief... is that nuclear energy is barely worth the candle... unless you develop the breeder,' while Glenn Seaborg, the head of the US Atomic Energy Commission, said in 1961 that a breeder reactor is 'the nuclear Goddess of Fertility, so fecund that, while producing power in plenty, it also breeds more fissionable material than it consumes'.[2] Plutonium was obtained through the reprocessing of spent fuels from the Magnox nuclear stations, and the amount acquired was magnified by the technical decision to store spent fuels at nuclear power stations in water-filled ponds rather than in dry storage. From there they were sent to the B205 reprocessing plant at Windscale, which was constructed in the 1960s to serve the Magnox programme. Thus the simple decision in favour of wet storage committed Britain to reprocessing on a scale that went beyond that required for the military programme.

This decision to reprocess nuclear fuels was made to appear economic partly through assigning a monetary value to plutonium, and partly by minimizing future financial provisions for nuclear waste disposal and decommissioning. As William Walker, a nuclear historian, remarks in his book *Nuclear Entrapment*, reprocessing 'was predicated on the belief that a large demand for plutonium would emerge and that the separated wastes would be treated and sites opened up for their disposal underground'. Reprocessing, while not affecting the amount of radioactivity, greatly increases the variety and volume of contaminated materials and equipment. Most of the radioactivity left by reprocessing is contained in concentrated high-level wastes (HLW), but a significant proportion is in intermediate and low-level wastes. Thus HLW creates a major (long-term) disposal problem, which still has not found a solution.

The commitment to reprocessing was further strengthened in the late 1960s by the failure of the British Magnox and AGR reactor designs – unlike the American Light Water Reactor (LWR) designs – to win export orders. The only commercial opportunity left was in the reprocessing market, and it was realized that a dedicated plant would have to be constructed if Britain were to become a major player in the expanding market for reprocessing services. Hence the decision in the early 1970s by BNFL to construct THORP at Windscale, and by mid-1975 contracts to reprocess 1500 tonnes of oxide fuel had already been negotiated with utilities from Japan, Germany, Sweden, Switzerland and Spain in the expectation that a new plant would be built.

As William Walker remarks:

> That all spent fuel from power reactors should be reprocessed, and that reprocessing capacities at Sellafield should be modernised and expanded to serve domestic and international markets, therefore became central tenets of British nuclear policy in the 1960s. As yet, they were uncontested. Indeed, reprocessing became more than a policy in this period: its pursuit became doctrine within British government and industry, an unquestioned amalgam of practices, routines, and policies founded on a confluence of interests and expectations. A shift in policy became unthinkable and unmentionable.

Furthermore the Japanese, who were fervent advocates of the FBR, were so eager to have their fuels reprocessed that they were willing to sign contracts highly favourable to BNFL, and provide much of the capital needed for the construction of THORP. Thus the support of the British government and media's early enthusiasm for THORP rested on the perception that reprocessing gave easy profits and large foreign earnings at essentially no risk. It was taken for granted that THORP would be approved through the local planning processes without recourse to public inquiry and with little public comment. After all there was no precedent for holding a public inquiry into a construction project at Windscale, and there was no public discussion, for example, of plans to construct the B205 Magnox reprocessing plant.

However a series of mishaps, beginning in 1973 with problems in a spent fuel pond which resulted in large discharges of radioactivity into the Irish Sea, provided unwelcome publicity for Windscale's activities. Local concern was further aroused by BNFL's planning application, which gave little detail about THORP's scale, purpose and consequences. The Labour government, although supportive of THORP and keen to avoid an inquiry, was increasingly wary of giving its consent without having at least gone through the motions of a public debate. This need for a public debate on nuclear energy was of particular interest to Tony Benn, the minister of technology, who according to William Walker 'did not hide his scepticism about the plans for THORP and his mistrust of its advocates'.

Then in October 1975 a front page story on Britain's 'Nuclear Dustbin' in the *Daily Mirror* took Benn at his word and made THORP part of his 'great energy debate'. Despite rising public concern the government was keen to push THORP forward and in March 1976 made a 'private' decision, apparently endorsed by the cabinet, in favour of THORP and of accepting foreign spent fuels. This was subject to one new condition, that the plutonium and radioactive wastes produced by reprocessing would have to be returned to their countries of origin. The government throughout the year tried to play down calls for an inquiry. Peter Shore, the secretary of state for the environment, dismissed requests in parliament in November1975 for an inquiry, and argued that local opinion was overwhelmingly in favour. However a U-turn came in late December 1976 after it emerged that the government had not been informed by BNFL of a further leak of radioactivity. After that Shore could no longer resist the clamour for a public inquiry, and this was held from June to October 1977 under Mr. Justice Parker (see Chapter 16).

This inquiry was not only the battleground but also the recruiting ground for the fledgling British anti-nuclear power movement. It also marked the turning point in the environment movement from the generally pro-nuclear stance of the old conservationists of the 1960s to the now prevalent anti-nuclear views of modern environmental activism as typified by FoE and Greenpeace. This book traces that development and the impact of this inquiry on the environmental movement.

Chapter 3

Twentieth-century
nuclear visions

The atomic age began in the early twentieth century, probably with the promotion of atomic energy by Frederick Soddy in *The Interpretation of Radium* (1909). Soddy was a leading nuclear physicist as well as a popular author and public lecturer, and he began the process of mythicizing atomic energy as an inexhaustible supply of power, which could be used to transform society, and heralding the possibility of an atomic utopia. But Soddy also gave us the image of atomic dystopia, warning us that a single mistake might not mean the smiling Garden of Eden but extremely disastrous consequences for mankind, perhaps a return to conditions of the stone age or even worse.

Soddy's views on atomic energy had a great impact on the public, and inspired H. G. Wells to write his famous science fiction novel *The World Set Free* (subtitled *A Story of Mankind*) published in 1914. Wells explicitly dedicated his book to Soddy's *The Interpretation of Radium*, and gave the world the first vision of what he called the 'atomic bomb'. He presents a very bleak picture of the horrors of atomic warfare with hundreds of cities destroyed but later the atom is used for peaceful purposes and it concludes with a cultural renaissance spawned by atomic power. This novel was highly influential in giving us conflicting images of atomic power, which were utilized by later science fiction writers. As science historian George Basalla concluded, in his essay *Some Persistent Energy Myths*, '*The World Set Free* might well stand as a symbol of the ambiguous response of Western men and women to atomic power. The choice was utopia or oblivion, the Garden of Eden or Armageddon'.

While other authors do not dispute the impact of Soddy's book and Wells' novel, they see the roots of nuclear euphoria (and ambiguity) dating back further to the discovery of X-rays and radium at the end of the nineteenth century. The discovery of the phenomenon of radioactivity in 1896, due to X-ray emissions from radium, led to a 'radium craze' amongst the public beginning in 1903. Radioactivity was seen as a scientific miracle with a wide range of positive health effects. Newspaper coverage in the early twentieth century was strongly positive, even though it was acknowledged that exposure to radium causes burns and eventual death. Scientists who died were considered 'martyrs to science', and the press, scientists and industry all promoted the view that the benefits of radium, particularly for 'curing' cancer, strongly outweighed any hazards. This aggressive support for radium and popular interest in all atomic discoveries must be seen in the context of the strong support for science and technology, which in the 1930s led to the utopian ideology of 'scientism', an unshakeable belief in the virtues of the scientific method. Stephen Hilgartner and his colleagues believed that the excessive belief in and promotion of atomic energy led to 'nukespeak', which they described as the language of the nuclear mindset – the worldview or system of beliefs of nuclear developers. They wrote in their book *Nukespeak*: 'Euphoric visions of nuclear technologies are an important expression of the nuclear mindset. The discovery of X-rays and radium at the end of the nineteenth century brought forward visions of a technological Garden of Eden. The philosopher's stone and the elixir of life had been found at last'.

The discovery of nuclear fission produced by U-235 in 1938 unleashed a torrent of similar imagery: nuclear-powered planes and automobiles would whisk us effortlessly around the globe, while unlimited nuclear electricity powered underground cities, farms, and factories. One example of this imagery was in a July 1940 article by Rudolph Langer in the popular US magazine *Colliers*. While Langer recognized the destructive possibilities of atomic energy, he swept them away by the argument that U-235 was too valuable to waste as an explosive and that 'industrious, powerful nations and clever, aggressive races can win at peace far more than could ever be won at war'. While his visions of nuclear-powered planes soaring to 50 miles above the surface of the earth and off-road automobiles roaming the planet were the stock of 1930s sci-fi fantasies, his social predictions of the end of poverty, class and war were the product of pre-war scientism and faith in social engineering, inspired by admiration for the Soviet Union and satirized in novels such as *Brave New World*. Langer boldly declared that we were about to enter 'a period of unparalleled richness and opportunities for all. Privilege and class distinction and other sources of social uneasiness and bitterness will become relics because things that make up the good life will be so abundant and inexpensive. War itself will become obsolete because of the disappearance of those economic stresses that immemorially have caused it'.

The main problems seen in this future are boredom caused by excessive consumption, uniformity of living standards and lack of novelty. While there may be 'eccentrics and criminals' who might want to use U-235 for 'destructive purposes', this problem Langer believed can be overcome by ensuring 'that the citizens of the future be educated better than they are now with respect to their social responsibilities'. If this is not sufficient then eugenics can be used, with Langer warning that certain kinds of abnormalities 'will have to be overcome or their possessors destroyed'. Finally Langer proposes a system of 'proliferation' control, saying that 'society will have to keep track of all the Uranium produced and refined, and take action at once against any individual who tries to accumulate a dangerous supply'.

Despite its utopian predictions and upbeat message, Langer's article unwittingly re-emphasizes Soddy's warning three decades earlier about the possibilities of nuclear destruction if sufficient care is not taken. How to exercise this care was to occupy the minds of the more thoughtful nuclear advocates for the next three decades, culminating in the famous (and much quoted) remarks made by Alvin Weinberg about a 'Faustian bargain' and the possibility of a 'nuclear priesthood' watching over nuclear wastes for generations to come. The most thoughtful and far-sighted of nuclear enthusiasts, Weinberg warned that the complex nature of nuclear power could lead to its alienation from the wider public. In an article in *Science* in July 1972, he wrote:

> It is a Faustian bargain that we strike: In return for this inexhaustible energy source, which we must have if we are to maintain ourselves at anything like our present numbers and our present state of affluence, we must commit ourselves – essentially forever – to exercise the vigilance and discipline necessary to keep our nuclear fires well behaved. As a nuclear technologist who has devoted his career to this quest for an infinite energy source, I believe the bargain is a good one.

However the price demanded of society for this 'magical energy source is both a vigilance and a longevity of our social institutions that we are quite unaccustomed to'. Weinberg warns of the need for effective nuclear waste disposal rather 'than keeping a priesthood that reworks the wastes or guards the vaults' similar to 'a military priesthood which guards against inadvertent use of nuclear weapons'.

The atomic bomb: Doomsday or paradise

After the dropping of the two atomic bombs on Japan in 1945 there was a rush by publishers to produce books on the new atomic age. The first was *The Atomic Age Opens* by Donald Geddes, a 256 page paperback compendium of news stories, editorials and pronouncements by world leaders, which along with other books, such as *Atomic Energy in the Coming Era* by David Dietz, gave a detailed description of the coming atomic utopia. The atomic bomb caused intense public debate in the United States on the morality of using atomic weapons, and fear and anxiety about the consequences of atomic warfare. The public was presented with two conflicting visions of the atomic future, and journalists created the image of society at a nuclear crossroads. However this public anxiety faded in a year or two to be largely replaced with a post-war nuclear euphoria. David Lilienthal, the first chairman of the Atomic Energy Commission (AEC) created in 1946, helped popularize this optimism, writing in *The Nation* in July 1946 of 'the almost limitless beneficial applications of atomic energy' while in the mass circulation magazine *Newsweek* they wrote about the 'miraculous powers of atomic-fission energy'.

William Laurence, a well-known writer for the *New York Times*, was an enthusiastic promoter of what Hilgartner calls the 'the sunny side of the atom'. In an article in May 1948 for the mass circulation *Woman's Home Companion* entitled 'Paradise or Doomsday', Laurence wrote in utopian tones, saying that humanity had, with its knowledge of how to harness the enormous power of the atom for good or for evil, been 'given a chance to enter a New Eden' and that atomic energy gave man a most powerful tool for 'abolishing disease and poverty, anxiety and fear'. Furthermore, Laurence enthused, provided man managed to avert the disaster that could transform him and all his works into 'a cloud of dust', he had in atomic energy 'a very philosopher's stone' with which he could 'remold the world nearer to his heart's desire'. As Hilgartner comments in *Nukespeak*, nuclear advocates after 1945 downplayed economic and proliferation problems:

> Once atomic energy was applied for peaceful purposes, a nuclear powered *paradise* would be at hand. Electricity *'too cheap to meter'* would power the new Golden Age. No problem would be too difficult to solve quickly and economically. Reactors would operate safely; the disposal of radioactive waste would be a nonproblem; international safeguards would allow the benefits of the peaceful atom to spread to every nation while preventing the spread of nuclear bombs.

However, as the coming atomic utopia failed to be realized, by the early 1950s there was increasing public skepticism – despite an optimistic façade at its chief promoter the AEC. In private many scientists were pessimistic. Robert Oppenheimer, the creator of the atomic bomb and post-war 'atomic hero', wrote in a report dated July 1947 to the AEC that 'it does not appear hopeful to use natural uranium directly as an adequate source of fuel for atomic power'. By the early 1950s, no AEC statement could conceal the basic fact that the era of atomic power was not arriving quickly, because of the reluctance of utilities to invest in a completely untried technology.

As Daniel Ford remarked in *The Cult of the Atom* (1982): 'The euphoria of the early postwar years gave way to gloom about the prospects of an atomic future' and leading magazines such as *Newsweek* wrote at the end of 1950: 'As yet no deserts have burst into bloom, no polar icecaps have melted away. The brave new world of cheap and abundant atomic power seems far more remote than it did just after the war'.

Atoms for Peace

In the early 1950s the dream of atomic energy had stalled, despite popular enthusiasm. Some economists openly expressed doubts as to its economic feasibility and practicality, and there was little immediate prospect and much disagreement on any likely timetable for the construction of a commercial nuclear power station. What revitalized the atomic dream and launched a commercial programme was the Atoms for Peace speech by President Eisenhower on 8 December 1953 at the UN. It marked a major shift in government atomic policy, ending the government monopoly on nuclear power, and rekindling the idea of a nuclear utopia. He urged that nuclear materials be used for peaceful purposes and to 'provide abundant electrical energy in the power-starved areas of the world' and stated: 'The United States knows that peaceful power from atomic energy is no dream of the future. That capability, already proved, is here – now, today'.

Once again the atomic visionaries rushed into print repeating the old 1940 predictions with a few new ones. There would be nuclear-powered planes, trains, ships and rockets; nuclear energy would genetically alter crops and preserve grains and fish; and nuclear reactors would generate very cheap electricity. New was the idea of using nuclear explosive to alter the landscape – a scheme to achieve notoriety as 'Project Ploughshare'.

The most famous phrase of this era, and one that was to haunt the nuclear industry for evermore, was uttered by Lewis Strauss, a former Wall Street investment banker and the new chairman of the AEC, in his speech to the National Association of Science Writers on 16 September 1954. Strauss set out the themes that the AEC wanted the media to present to the public. Electric power from the atom, he said, could be available, according to the AEC's experts, in 'from five to fifteen years, depending on the vigor of the development effort'. He then went on to give his vision of the atomic utopia, saying: 'It is not too much to expect that our children will enjoy in their homes electrical energy too cheap to meter, will know of great periodic regional famines in the world only as matters of history, will travel effortlessly over the seas and under them and through the air with a minimum of danger and at great speeds, and will experience a life span far longer than ours… This is the forecast for an age of peace'.

The phrase 'too cheap to meter' has become the cliché of the nuclear age and the stick with which the anti-nuclear campaigners from 1970 continually reminded the public about the failed promises of the nuclear promoters. However Strauss was only repeating the dream of atomic protagonists since 1940, when Rudolph Langer first said that in the forthcoming atomic age 'energy has become so cheap that it isn't worth making a charge for'. The year before Strauss' speech, Ralph Lapp in his book *The New Force* had repeated an earlier prediction by Robert Hutchins that 'heat, light and power… will be so cheap that their cost can hardly be reckoned'.

In his speech Lewis Strauss invited his audience of science writers to 'work together' with the AEC and its scientists to educate the public about the atom and its promise. As Daniel Ford commented: 'From the laudatory articles on nuclear energy that appeared over the next two decades – and the rarity of any critical coverage of the potential hazards – it is evident that the national media responded to the chairman's invitation as he had intended'.

Thus with unquestioning support from the media, and unqualified endorsement by Congress and the administration, the advocates of a large nuclear power programme proceeded, unchallenged, with their ambitious enterprise. They were aided by the stream of euphoric books, with titles such as *Our Friend the Atom* by Heinz Haber and *The Fabulous Future* edited by *Fortune* magazine. There were even films, such as *The Genie and the Bottle*, a 1956 book turned into a Walt Disney film.

The Fabulous Future (1956) is subtitled 'America in 1980' and has a series of essays on what life in America would be like, written by leading figures in US political and social life, including Earl Warren, Adlai Stevenson and John von Neumann, then a board member of the AEC. Unlike his fellow contributor, the industrialist David Sarnoff, who stated that 'it can be taken for granted that before 1980 – ships, aircraft, locomotives and even automobiles will be atomically fuelled', von Neumann is less optimistic about atomic technology and sees its destructive side. In his essay, *Can We Survive Technology?*, he devoted one of his fifteen pages to nuclear power, saying that the future lay not with steam-generating fission reactors but with exploitation of 'still more abundant modes' (an oblique reference to the fast breeder or the possibilities of fusion energy). Given steady development he believed that 'in a few decades hence energy may be free – just like unmetered air'.

In this technologically optimistic era there were few people who publicly raised the question of the harmful consequences of the peaceful uses of atomic energy, such as radiation hazards and waste disposal. Scientists knew of these adverse consequences right from the early years of the twentieth century, during the era of the 'radium craze'. As early as May 1940 a report of the National Academy of Sciences talked of 'radioactive poisons' as products of atomic fission. As George Basalla commented: 'In all this euphoria there were few who discussed radioactive wastes, the dangers of radiation, or the possibility of a dangerous accident at a nuclear generating plant... All these issues had been raised publicly by scientists, but layman and nuclear expert alike were all caught in the grips of a myth that portrayed atomic energy as perfect, inexhaustible, and utopian in its social implication'.

Nuclear power comes on line

After Eisenhower's Atoms for Peace speech there was rapid commercial development of nuclear power. Construction started on the first reactor in September 1954 at Shippingport and on 18 July 1955 the reactor came on line. Following this success the first UN Conference on the Peaceful Uses of Atomic Energy was held, in Geneva in August 1955. The pace of commercial nuclear power development quickened from December 1963 when General Electric offered a fixed price (turnkey) contract to build the Oyster Creek power station. By the end of 1966 another eleven turnkey contracts were signed together with 16 other contracts.

The rapid expansion of nuclear power in the USA in the 1960s was, according to nuclear historian Steven Cohn in his book *Too Cheap to Meter*, partly due to the public's buoyant faith in business and public leaders and their faith in the ability of science and technology to unambiguously solve social and economic problems. Even those who expressed criticism of American society and its political structures saw nuclear power as a liberating influence. For instance in 1962 the founding document, *The Port Huron Statement*, of the Students for a Democratic Society (SDS) declared: 'Our monster cities, based historically on the need for mass labor, might now be humanized, broken into smaller communities, powered by nuclear energy'. By the mid-1960s with a large nuclear programme underway the dream of atomic energy and an all-electric future seemed imminent. A 1967 Westinghouse pamphlet entitled *Infinite Energy* declared that nuclear fission 'will gives us all the power we need and more' and: 'That's what it's all about. Power seemingly without end. Power to do everything man is destined to do. We have found what might be called *perpetual youth*... the hope and exuberance of *boundless energy*'.

The illustration of the future in this pamphlet gave a vision of the all-electric future, a world very similar to that drawn in Langer's article in 1940, of homes with movable roofs, indoor farming, and computer-operated agricultural machines. Nuclear advocates pointed to graphs showing falling nuclear costs and the continued historic decline in electricity prices, and were still in the late 1960s boldly proclaiming that such trends would lead to the fulfilment of the age-old dream of electricity too cheap to meter. As the US anti-nuclear movement gathered pace in the late 1960s and early 1970s, the nuclear vision remained undimmed in the eyes of its advocates. In 1971 Glenn Seaborg, the head of the AEC from 1961-71, and William Corliss published their book *Man and the Atom*, subtitled *Building a New World through Nuclear Technology*. In predictions for 2000 they forecast that nuclear power would be a phenomenal success, bringing unimagined benefits for the greater part of humanity. Seaborg believed that the future of civilization was in the hands of the nuclear scientists who formed the elite team that would build a new world through nuclear technology. The nuclear dream was still the same as three decades ago with Ford commenting on Seaborg's vision:

> The deserts could be made to bloom, sea water could be made potable, mountains could be moved, rivers diverted – all as a result, he prophesied, of 'planetary engineering' made possible by the miraculous new element that he had discovered. There could be vast farming and manufacturing centres, or 'agro-industrial complexes', built around giant nuclear electric-generating stations – each nuclear power plant surrounded by its own little Eden.

Seaborg was an enthusiastic promoter of the use of nuclear bombs to reshape the landscape and alter nature, a scheme that had official US government backing in the form of Project Ploughshare. Seaborg wrote: 'All of humanity's efforts to restore the Garden of Eden have been futile so far. Man's machines have not been powerful enough to compete with the forces of nature'.

The end of the nuclear dream

The fall from grace of nuclear power had more to do with the perceived shortcoming of the nuclear bureaucracy than with technical failings. In the USA, the Atomic Energy Commission, and in Britain, the AEA, were most probably no worse than any other bureaucracy, it was just that the public had such high expectations of the nuclear endeavour after decades of propaganda. When the nuclear utopianism collided with the grim realities of the 1973 'energy crisis', of power shortages and petrol queues, disillusionment rapidly set in as it had in the late 1940s. This time the factors that had restored confidence in the mid-1950s were under attack, namely public trust in business and government leaders and faith in the ability of science and technology to solve social and economic problems. The problems with nuclear power were as always, but now the public no longer had faith in the ability of government, business and science to solve them. Since the early years of the twentieth century there had always been ambivalence amongst the public about the benefits of atomic energy and leadership by technocrats, and these cultural countercurrents, submerged during the post-war era, resurfaced in late 1960s. As Steven Cohn commented, 'cultural ambivalence towards nuclear energy quietly persisted, however, emerging publicly in the late 1960s and early 1970s when a different social context permitted bundles of cultural beliefs to reach critical mass and capture social influence'.

The nuclear advocates claimed that their critics were unduly pessimistic, and were acting as prophets of doom. As early as 1955 Fred Cottrell, discussing speculation on the impact of atomic power, wrote in his book *Energy and Society*: 'The prophets of doom say it will condemn man to a return to the cave' but – more optimistically – 'there are those who see in atomic energy the basis for a brave new world, in which war is impossible and men are freed from the necessity to work'. He saw the introduction of new energy sources always giving rise to this speculation of doom or utopia, and concluded that 'man has only abandoned one source of energy for another because of the prospect of overwhelming advantage, and then with reluctance'. So what of the dreams of the atomic age bringing unimagined benefits for most of humanity? Vaclav Smil concluded in his book *Energies* (1993) that 'none of this has come to pass, and the retreat of nuclear generation has been almost certainly the costliest technical miscalculation of the twentieth century'. Hilgartner lays the blame for the failure of nuclear power on the 'nuclear mindset' of its promoters, arguing that:

> Time and time again, nuclear developers have confused their hopes with reality, publicly presented their expectations and assumptions as facts, covered up damaging information, harassed and fired scientists who disagreed with established policy, refused to recognize the existence of problems, called their critics mentally ill, generated false or misleading statistics to bolster their assertions, failed to learn from their mistakes, and claimed that there was no choice but to follow their policies.

A more measured response is from Steven Cohn who sees the tragedy behind an idealistic venture, and he ended his book *Too Cheap to Meter* (1997) with the words:

> The sad conclusion from my perspective is that the nuclear dream has not worked out. The technology has failed and should be put aside until other energy options

have been exhausted and the industrial subculture that nurtured the first nuclear era dismantled. I find this a sad conclusion because the nuclear dream was compelling, the imaginations behind it were talented, and the human energy and economic wealth mobilized to pursue it were enormous.

SF visions

Science fiction writers often explore the potential implications of new technologies and this was very much the case with nuclear power. Indeed, in some ways they set the tone for the public responses that occurred later, arguably to a much greater extent and earlier than other media. Writers such as Paul Boyer and Spencer Weart believe that SF has been a powerful force in shaping societal attitudes to the atomic bomb. Given the close links between the bomb and nuclear reactors, SF has also been influential in shaping attitudes to nuclear power. The two were first linked by Frederick Soddy, who gave us the twin visions of atomic energy as both a source of abundant energy and of immense destruction, both of which drew on early literary images of a scientific future.

Wells in *The World Set Free* predicted that in the 1930s scientists would release atomic energy, and that by the 1950s this knowledge, in the form of what he called 'atomic bombs', would lead to a devastating worldwide nuclear war, with hundreds of cities left in ruins. Technically it was highly inaccurate, as the atomic bombs were two-foot black spheres, dropped by hand from open aircraft, which went on exploding forever, though their power was halved every seventeen days, leaving gold as a by-product. Despite the mushroom cloud – our quintessential image of the atomic bomb – being absent, the essence of the bomb, its name and the terror it conveys are definitely foreshadowed. The atomic bomb (or a weapon of similar destructive power) soon became a staple of SF, which reflected the vigour of two of the twentieth century's favourite nightmares – -the fear that the human race has an unlimited capacity for self-destruction and the dangers of uncontrolled scientific experimentation. However others saw optimism in *The World Set Free*, as the novel ends with the formation of a world government, which sets the world free from war, insecurity and the struggle for existence. As Spencer Weart commented in *Nuclear Fear*:

> At the story's end citizens could travel where they chose in atom-powered aircars, building atom-powered garden cities in deserts and arctic wastes, enjoying liberty and free love. Wells had neatly fitted together fragmentary notions about science and atomic energy to craft the first full-scale scientific legend of atomic Armageddon and millennium.

Alvin Weinberg was one youthful scientist who was deeply influenced by the atomic visions of Wells, to whom he later paid tribute. In 1966 Weinberg predicted a glorious future for nuclear power, saying that 'if nuclear energy does not, as H. G. Wells put it, create "A World Set Free", it will nevertheless affect much of the economy of the coming generation'. Wells' imagery of atomic-powered cities was evident in the plans in the 1960s by scientists at Oak Ridge, the 'City of the Atom' in Tennessee, to develop the 'nuplex' (a nuclear complex). This was to be a town centred on reactors, which could be located anywhere in the world regardless of geography or climate. Alvin Weinberg proudly explained in 1971 in *Foreign Affairs* that this was precisely the dream of a world set free that he had learned from H. G. Wells.

However public ambiguity about atomic energy was reflected in inter-war SF novels: for some it was a means to disaster, but to others it was a means of salvation. One early eco-catastrophe novel, *Nordenholts Million* by J. J. Connington (1923), had disaster only averted by the heroic development of an atomic energy machine. The treatment of atomic energy by SF writers before the 1940s was in the long tradition that the civilization of the future could not progress technologically without being in possession of a plentiful and efficient source of power. The discovery of 'atomic energy' in the early 1900s gave SF writers a new source of unlimited power, which could be used in their stories. Atomic energy was 'clearly magic in disguise', ranking along with such other imaginary powers sources as 'vril' (used by Bulwer Lytton, in 1871, in *The Coming Race*) and 'apergy' (by Percy Greg, in 1880, in *Across the Zodiac*). By the 1920s 'atomic energy' had become the power system which would provide limitless energy and bring utopia. A typical example of such banality is *John Sagar* (1921) by 'Nedram' where an inventor gains control of atomic energy, becomes self-proclaimed 'Master of the World' and sets about creating a utopia for all. In the late 1920s Hugo Gernsback, the editor of *Amazing Stories*, vigorously promoted atomic energy in his magazine. He predicted its use in the near future, and referred to the coming era of high technology as 'The Atom-Electronic Age' or 'The Age of Power-Freedom'.

Atomic power

In the 1930s SF stories dealing with atomic power became much more realistic, primarily due to the influence of John W. Campbell, the editor of *Astounding Science-Fiction* (ASF). This magazine was widely read by science and engineering graduates and prided itself on its scientific accuracy. Campbell, who had thought deeply about the realities of atomic energy, had published a number of stories since 1930 – *When the Atoms Failed* (1930) and *Atomic Power* (1934) – and in editorials from June 1938 had declared the reality of atomic fission. The announcement of the achievement of atomic fission by Niels Bohr in January 1939 was followed by widespread press coverage and speculation on its uses, and enhanced credibility for Campbell. In September he published Robert Heinlein's first atomic energy story *Blowups Happen*, with an editorial announcing that this story was based on the latest discoveries. This is perhaps the first SF story on the possibility of an accident at a nuclear power station, and is noteworthy in the way it deals with the psychological stress involved in working at a nuclear power station when one mistake might cause widespread devastation. This theme of the potential for a disastrous accident was again explored in *Nerves* by Lester del Rey, published in ASF in 1942.

Post-war pessimism

After the dropping of the atomic bomb in 1945 SF writers were acknowledged as prophets proven right, and many managed to become professional writers. Some, such as Ray Bradbury and Robert Heinlein, prospered as the mass circulation magazines *Colliers* and *The Saturday Evening Post* began to publish SF. Others became writers of popular science and academic consultants, but most had an ambivalent position on atomic energy because of the bomb. A few maintained their optimism, with Theodore Sturgeon writing in *ASF* in December 1945 that he 'celebrated the possibilities of nuclear power for changing the world'.[3]

In the immediate post-war era while science journalists were producing exalted prophecies of the bright promise of atomic energy, SF writers were almost invariably producing bleak and pessimistic stories on its social consequences. But as cultural historian Paul Boyer remarked in his book *By the Bomb's Early Light,* 'these two responses that seemed so contradictory – the terror of atomic war and the vision of an atomic utopia – were in fact completely interwoven'. The atomic bomb in fact ended the SF vision of atomic energy as the route to utopia; it was now the road to dystopia. As the initial flush of enthusiasm for atomic energy by science journalists faded by the late 1940s under criticism from economists and scientists, the cultural climate turned deeply pessimistic and inward looking. Thus at the dawn of nuclear power there were deeply conflicting images of atomic energy, and of the future direction of society. The optimists, according to Boyer, believed that the atomic bomb could be controlled and 'atomic energy made a great blessing to mankind', while the pessimists, the SF writers, 'were offering a counter vision almost unrelieved in its bleakness and despair'. As SF historian I. F. Clarke commented in his book *The Pattern of Expectation*: 'For some twenty years, from the end of the 1940s to the late 1960s, the genre concerned itself with dystopian visions of despotism and disaster, and with the ruthless repudiations of twentieth century civilization

– parables in which a nobler species replaces Homo sapiens, myths of rebirth in which the survivors rebuild the good society'.

This was the age of the great dystopian novel: *1984* by George Orwell, *Ape and Essence* by Aldous Huxley, *Player Piano* by Kurt Vonnegut and *Fahrenheit 451* by Ray Bradbury. There were no longer grand schemes for building a utopian urban society; the emphasis was now on individual and community survival. No longer was there the pre-war belief amongst most SF writers (except Isaac Asimov and Arthur C. Clarke) in the benefits of technical progress, and this reflected post-war utopian thought that had to a large extent distanced itself from the idea of progress. Instead it focused on the idea of a retreat to a simpler life, and perhaps this was the only glimmer of hope available to writers of apocalyptic novels. It was making a virtue out of necessity: society could be reborn after the bomb. At the same time SF literature was rediscovering the pastoral and communitarian tradition that dated back to the late nineteenth century, with its intense dislike of modern industrial society. The search was now for the 'good life', one based more on moral values that technology and, in the words of I. F. Clarke, post-war utopian fiction aimed 'to redeem the moral failures of our times by revealing a society in which the individual is helped to lead a life that is good in itself and can become better'.

Along with fears of nuclear holocaust came, from the late 1940s, renewed Malthusian warnings on the imminence of world over-population, famine and resource shortages. This further reinforced the implicit ecological message of the few utopian writers – B. F. Skinner in *Walden Two* and Robert Graves in *Seven Days in New Crete* – on the need to return to small pastoral communities. To these utopians, and the few SF writers offering qualified messages of hope, the end of modern civilization was not lamented. For it was the chance for a new beginning, where the survivors might achieve an existence in harmony with nature and with each other. The post-war pessimism about the future was reflected in attitudes to radioactivity (from the atomic bomb), and the destruction of the environment by radioactive poisoning became one of the most horrifying aspects of the post-atomic scenario. These stories helped to bring about a much-increased sensitivity to the idea of insidious poisons in the environment. As SF writer Brian Stableford commented in *The Encyclopedia of Science Fiction*: 'It was in the early 1960s, however, that the problem was brought very sharply into focus, largely due to the publication of *Silent Spring* (1962) by Rachel Carson which argued that pollution of a radically new type had begun, involving nonbiodegradable substances which accumulated in living matter to fatal concentrations... Awareness of these threats was very rapidly absorbed into SF, and virtually overnight became a standard feature of near-future scenarios'.

One of the first SF authors to promote 'ecology' in their books was Frank Herbert in *Dune* (1965); here the planetary hero is called the 'Ecologist', and human survival depends upon fanatical water conservation. Herbert's novel *The Green Brain* (1966) is an SF 'Silent Spring' – on the dangers of attempting to exterminate insect life through pesticide use. From the mid-1960s the eco-catastrophe theme in SF became more common. These novels, by such authors as Harry Harrison, James Blish and John Brunner, were reinforced by the emerging popular eco-doom literature from authors including Garrett Hardin and Paul Ehrlich. SF was seen as a way of getting the 'eco-doom' message across (because many teenagers never read anything else). Both Hardin and Ehrlich wrote SF novels, and Ballantine published an SF anthology, *Voyages: Scenarios for a Ship called Earth*, for the Zero Population Growth movement in 1971.[4]

Chapter 4

Public opinion

The post-war ambiguity of SF writers about atomic energy was reflected in public opinion. There have been polls on nuclear power in most nations and at most times since 1945, and analysis of them by Spencer Weart reveals that there are five groups of roughly equal size:

> The first group were strong advocates, confident in the ability of experts to keep reactors safe and impressed by the benefits of economic growth. Directly denying all that and about equally numerous were the convinced opponents, at first silent but increasingly vocal. In between were those who leaned one way or another: a group who favoured reactors even while harbouring some misgivings about their safety, and a group who frankly feared reactors but were not convinced they should be banned. Both these middle groups would usually go along with whatever the authorities decided.

Finally the last group consisted of people with no position, and often no interest in the whole question. Thus only about 20% of the public could be called nuclear supporters, with the majority of about 60% either against or fearful, with the remainder as 'don't knows'. It was the two middle groups who were the 'swing' voters. When the authorities were confident they joined the strong advocates, when the authorities were divided over safety issues they hesitated and adopted a 'wait-and-see' attitude. Thus public confidence and trust in nuclear authorities was crucial in winning their support, and when this began to crumble in the early 1970s support fell away.

Advocates and opponents of nuclear power were spread quite evenly throughout most social groups, with education making little difference to their views. Studies showed that the way people felt about nuclear power was mostly independent of how much they knew about it – in fact, most people had only rudimentary knowledge mixed with various bits of misinformation. Weart concluded that 'when a person took a nuclear stance it was not from some special knowledge or lack of it, but as part of a total approach to society'. The only factor, according to the polls, associated with anti-nuclear opinion was sex: women in all countries, in all time periods, were consistently more anti-nuclear than men. Weart's explanation for this divergence in opinion lies in the symbolism of nuclear technology for women, and that: 'It had become most specifically associated with aggressive masculine imagery: weapons, mysteriously powerful machines, domination of nature, contamination verging on rape. On no other technological issue was the sexual imagery so thoroughly developed and, from a woman's standpoint, so viscerally disturbing'.

Public attitudes to nuclear power can also be surmised from analysis of the titles of articles published in popular magazines. An analysis of *Readers Guide to Periodical Literature* revealed that in 1950 well under 10% of the titles had suggested that there was anything to worry about. Anxiety began rising in the 1950s, paused during a period of indifference to nuclear power during the mid-1960s and rose steadily thereafter, reaching a plateau of about 25% in the early 1970s, before rising from 1975 to over a third by 1980, and surging upwards to 90% by 1986. In contrast the optimistic articles peaked at nearly 50% in the mid-1950s under the influence of the Atoms for Peace rhetoric, then began a steady decline from the early 1960s reaching less than 10% by 1970 and stabilizing there. The crossover point from optimism to pessimism is the late 1960s, the start of the anti-nuclear power movement in the United States, and also the time of minimal interest by periodicals in nuclear power issues. A similar result was obtained in an analysis of books on nuclear energy in the Ohio state library; of the 77 books published up to 1965, only 5 were anti-nuclear with 22 pro- and 50 neutral. However between 1965 and 1980 of the 83 books published, 45 were anti-nuclear with only 18 pro- and 20 neutral.

Weart's 'nuclear fear'

Spencer Weart uses the term 'nuclear fear' to explain opposition to nuclear power, and in his opinion this fear was heavily influenced by images of the nuclear bomb in popular culture. The bomb had become a very powerful symbol of mass destruction, the apocalypse, and a world ruined by radiation. It had also become associated with myths of 'technology out of control', of 'mad and bad' scientists experimenting with things 'best left unknown', and of control of nature and mankind by a remote elite. Atomic energy had become the cliché of every SF horror story: of mutant insects caused by radiation, of death rays created by the evil scientist bent on world control, of the crazed military officer launching a

surprise nuclear strike, or of bureaucratic utilities ignoring nuclear safety problems. The emerging environmental movement was able (perhaps unexpectedly)
to capitalize on these highly emotional images when it got involved in local
campaigns against the building of nuclear power stations. Weart commented:
'Nuclear fear was by no means the only form of imagery, let alone the only social
force, behind the environmental movement. But nuclear fear took a special place.
It raised emotions earlier and on a more visceral level than any other issue. And
it served as a banner that could rally everyone around'.

Early analysts of the US anti-nuclear movement recognized this nuclear fear
and argued that the concerns that induced opposition to a nuclear power station
were based on more than simply environmental issues. For instance Ebbin and
Kasper in their book *Citizen Groups and the Nuclear Power Controversy* (1974)
remarked: 'Fear of things nuclear or atomic is not uncommon among the population in general and is rather evident among intervenor groups in particular: fear
of nuclear accidents, fear of radiation exposure, fear of propinquity to nuclear
plants, fear of long-range unknowns, fear of technological error and commitment
to a technology imperfectly understood'.

They remarked that this fear had an evident historical foundation, as most
adult Americans were first made aware of nuclear power with the dropping of the
atomic bomb. The image of the mushroom-shaped cloud, they concluded,
'remains clear and frightening in the minds of the American public'.

The WHO study group

This nuclear fear was recognized early on by the nuclear authorities, and in
1957 the World Health Organization (WHO) set up a study group on Mental
Health Aspects of the Peaceful Uses of Atomic Energy, and an account of its
workings is given by one of its members, Ritchie Calder, then a science journalist.[5] Calder wrote in his book *Living with the Atom* that the study group sought
to change the 'irrational and deplorable' fear of science meddling with things that
would be better left alone, or in other words 'to re-enthrone reason and rebut the
New Superstition'. The WHO study group made clear that by 'mental' it meant
'emotional' and Calder wrote that to it 'certain disturbing features became plain.
One was the universal disquiet about atomic energy, in terms not only of its
potentialities for destruction in a nuclear war but of its peacetime implications'.
They began to realize that 'civilized man tends to cower, like his Neanderthal
forefathers, in the dark cave of his own emotions. We are back in the "childhood
of mankind". Man's anxiety about his own search for knowledge and for power
is reflected almost universally in myth and legend and still lurks in our own nature
today'.

These myths and legends, such as those of Pandora and Faust, were renewed in SF films and novels of the 1950s, in which radiation from the atomic bomb created mutated creatures that would wreak destruction on mankind (as in the highly popular 1954 film *Them!* about giant killer ants). The WHO study group sought to explain this fear 'that trespassing in the Unknown will invite a kind of cosmic revenge on mankind' in psychological terms as the tendency to relapse into more primitive forms of thought and feeling known as 'regression'. As Calder reported, 'The psychologists also have an explanation for the universal fear of "fallout" and atomic waste. These are associated with feeding and excretion. The danger to food is generally the most disquieting concern about fallout or the risks of a nuclear mishap and, so the psychologists say, there is a symbolic association between atomic waste and body waste'. This fear of contamination of food was used by opponents of nuclear bomb testing in the late 1950s, when they campaigned on the issue of radioactive strontium-90 in milk. This fallout threat to milk was also used in 1963 by campaigners against a nuclear reactor at Bodega Bay in California, an issue that particularly appealed to women. Such campaigns exploited deep-seated myths of milk as the most sacred of food. As Weart observed, since the time of witchcraft trials accusations of obscene attacks on milk have been a powerful weapons against enemies.[6] Weart also agrees on the psychological importance of atomic wastes being viewed as excrement suggesting that, on every level of human thought, radioactive wastes – in association with weapons – were seen as 'filthy insults against the proper order of things'.

Linking bombs and nukes

The linkage in the public mind between atomic bombs and power had always been strong, and anti-nuclear power campaigning could be seen as a substitute for the failed earlier attempts to rid the world of nuclear weapons. The hostility to nuclear weapons sought an easier outlet: what psychologists would call 'displacement', a hostility that shrinks away from what is too threatening, directing itself onto some other target instead. This can be considered as a rational response to an insoluble problem, just as the 1950s campaign by Linus Pauling and others against atomic fallout hid their primary goal of reducing the likelihood of the use of atomic bombs.

David Lilienthal, a veteran of the attempt to control the arms race, remarked in his book *Atomic Energy: A New Start* (1980) that people in the 1970s were attacking reactors as a 'surrogate for bombs'. Weart comments that the fear and hostility previously directed towards bombs was now directed onto nuclear power, and this displacement became clear to him when one person remarked to him 'I can't do anything about the bomb, but I can do something about reactors'.

Initially in the 1970s the anti-bomb (or peace) movement and the anti-nuclear power movement were distinct, due to differences of style, tactics and constituencies. The peace movement argued that the problems of nuclear power were simply trivial compared to the dangers of nuclear war, while the anti-nuclear movement was reluctant to compromise its 'political neutrality' by campaigning on an issue that had strong pacifist and left links. In the UK most campaigners did not link nuclear power to nuclear weapons until the early 1980s when hundreds of local groups networked in the Anti-Nuclear Campaign (ANC) joined together with a reinvigorated CND to oppose both nuclear weapons and nuclear power.

Mastery of nature

The nuclear power reactor, to both supporters and opponents of technological progress, stood as a 'full symbolic representation of attempts by the forces of order to master nature'. Weart goes on to explain:

> That symbolism can explain better than anything else why nuclear reactors were singled out, far more than any other technology, for utopian hopes on the one hand, fear and hostility on the other. In terms of hierarchical control by a technical elite, reactors were not obviously worse than, for example, the telecommunications industry; in terms of prospective hazards to our daily lives, reactors were not obviously worse than, for example, the chemicals industry. What reactors did offer that nothing else could match was a unitary image tying together everything involved in the battle between 'nature' and 'culture'.

Environmentalists were at the heart of this battle against the mastery of 'nature', and thus the nuclear reactor became the prime symbol of this largely idealistic battle. Also reactors, as large objects which people could see as 'blots on the landscape', served as a very convenient concrete reality in this ideological dispute about the direction and control of society. Thus the anti-nuclear power campaign was able to link local rural 'NIMBY' protests about the construction of large industrial projects with urban romanticism about the destruction of nature by technological elites.

Chapter 5

Energy utopianism

One person's utopia is, as many commentators have pointed out, another person's hell. The same dichotomy is also true for new energy technologies, which tend to be launched on the basis of zealous commitments and grand projections of changing society. This tendency to attach utopian attributes to new technologies has been widespread particularly in American society. One American writer on technology, Langdon Winner, remarked in his book *The Whale and the Reactor* that it is a recurring fantasy in industrial society that a new technology will end people's alienation from the current centralized political and technological system. Winner wrote: 'Dreams of instant liberation from centralized social control have accompanied virtually every important new technological system introduced during the past century and a half. The emancipation proposed by decentralist philosophers as a deliberate goal requiring long, arduous social struggle has been upheld by technological optimists as a condition to be realized simply by adopting a new gadget'.

Winner talks about a great tradition of optimistic technophilia and remarked: 'It is not uncommon for the advent of a new technology to provide an occasion for flights of utopian fancy. During the last two centuries the factory system, railroads, telephone, electricity, automobile, airplane, radio, television and nuclear power have all figured prominently in the belief that a new and glorious age was about to begin'.

This belief in the liberating power of new technologies was common amongst intellectuals and writers in the 1930s. Cultural critic Lewis Mumford believed that there could be a return to nature by way of modern technology, and in his 1938 book, *The Culture of Cities*, he 'waxes lyrically about the liberating qualities of electric power lines, aluminum, radios, automobiles, superhighways and planes'. This technological utopianism is beyond the scope of this book but is discussed in great detail by Howard Segal in his book *Technological Utopianism and American Culture*

(1985) where he examines the historical phenomenon of an uncritical faith in technology's ability to solve all problems.

It is therefore no surprise that supporters of new energy technologies, from the steam-engine to the solar cell, have since the mid-nineteenth century proclaimed that these technologies could act as a motive force for desired social change and as a means of returning to a 'cherished naturalistic bliss'. The discovery of electricity, in particular, gave birth to ideas of a new bond between nature and society termed 'the rhetoric of the electrical sublime'; ideas used by late Victorian writers, such as reformers (as in Howard's *Garden Cities of Tomorrow*), by revolutionaries (as in Kropotkin's *Fields, Factories and Workshops Tomorrow*) and in utopian science fiction (as in Bellamy's *Looking Backward*). These ideas and images for new energy technologies were termed 'energy utopianism' by George Basalla, and for electricity they were termed 'electricity triumphalism' by Bill Luckin in Britain and the 'mythos of the Electronic Revolution' by James Carey and John Quirk in the USA.

Basalla's 'energy utopianism'

In his 1982 essay *Some Persistent Energy Myths* Basalla reviewed the history of the introduction of a new source of energy into society over the last two centuries. He remarked that there was a recurrent energy myth, and 'any newly discovered source of energy is assumed to be without faults, infinitely abundant, and to have the potential to effect utopian change in society. These myths persist until a new energy source is developed to the point that its drawbacks become apparent and the failure to establish a utopian society must be reluctantly admitted'. Social commentators, he believed, do not seem to learn from previous mistakes by over-enthusiasts, for the next source is not handled in a more restrained fashion. Basalla commented: 'Instead, the recently discarded energy myths are resurrected and bestowed upon the newcomer. And so the cycle continues, with only the myths remaining unaltered while various sources of energy enter and leave the public spotlight'.

Basalla noted that these fantasies attached to energy use and production are shared by a wide section of society: technologists, capitalists, and the general public alike. Finally he concluded: 'Energy myths are particularly dangerous because they blind us to the realities of new energy sources by promising a golden land of the future and ignoring the real problems of today'.

Energy utopianism is also strongly linked to social utopianism, the desire to create new forms of society based on using the latest technology to liberate people from work or to create mass wealth. Thus the widespread application of new sources of energy often became a key part of the political ideology for those wishing to remake society.

Electro-propaganda

In the early twentieth century the provision of electricity, particularly to rural areas, was seen as a key component of the modernization and industrialization of society. This desire for electrification was perhaps most intense in the Soviet Union after the 1917 revolution, and Lenin enthusiastically declared in December 1920 at the Eighth All-Russia Congress of Soviets that: 'Communism is Soviet power plus the electrification of the whole country... only when the country has been electrified, and industry, agriculture and transport have been placed on the technical basis of modern large-scale industry, only then shall we be fully victorious'.

Lenin's view of the transforming merits of electrification found similar support in the United States. For instance, Joseph Hart, a professor of education, extolled in the US magazine *The Survey* (March 1924) on the liberation that electricity would bring:

> Centralization has claimed everything for a century: the results are apparent on every hand. But the reign of steam approaches its end: a new stage in the industrial revolution comes on. Electric power, breaking away from its servitude to steam, is becoming independent. Electricity is a decentralizing form of power: it runs out over distributing lines and subdivides to all the minutiae of life and need. Working with it, men feel the thrill of control and freedom once again.

These ideas were supported by Lewis Mumford who, together with other conservationists such as Gifford Pinchot, formed the Regional Planning Association of America around the goals of public electric power and community planning. These goals were adopted by the US government during the 1930s, with the creation of the Tennessee Valley Authority (TVA) and the Rural Electrification Administration. This intense admiration and enthusiasm for electrification as the harbinger of modernization was widespread in all western countries, especially those that saw the state as the leading agent for social change through widespread planning. Bill Luckin, writing in his book *Questions of Power* on the construction of the National Grid in Britain in the inter-war years, develops the concept of 'electricity triumphalism', a progressive ideology that would overcome traditionalist thinking on technical change. He remarked that this 'was grounded in the scientific premiss that economy, society and culture would be rapidly and radically transformed by the new source of energy. Forged by electrical engineers, contractors, salesmen, journalists and technocrats, this progressivist ideology was buttressed by powerful state bureaucracies'.

This rhetoric or 'electro-propaganda' from the technical press, such as *Electrical*

Review, promoted the image of the forthcoming 'golden age of electricity' to consumers, based on its association with 'progress', and with a modern, middle-class lifestyle based on health, leisure and domestic appliances. As Luckin commented:

> Electrical triumphalism was more than a generalized and at times quasi-religious rhetoric, growing out of and moulded by a deep defensive reaction to the existing technological order. It must also be seen as the shaper and bearer of images which made up the 'public face' of electricity and which persuaded large numbers of the uncommitted to invest, symbolically as much as economically, in the 'future' rather than the 'past'.

Critics

There were few critics of electro-propaganda in the post-war years. Perhaps the first were James Carey and John Quirk who examined the notion of electricity as a liberating force in two essays in *The American Scholar* in 1970. In the first they reviewed the futurist ethos presented by modern intellectuals, such as Marshall McLuhan and Buckminster Fuller, that 'identifies electricity and electrical power, electronics and cybernetics, with a new birth of community, decentralization, ecological balance and social harmony'. They then identified Harold Innis, the Canadian economist, as the first to produce a systematic critique of the notion that electricity would replace centralization in economics and politics with decentralization, democracy and a cultural revival. In fact, Innis argued, the trend of new technologies was towards increased territorial expansion, spatial control, commercialism and imperialism – what we would today call 'globalization'. As they remarked: 'He recognized that the speed and distance of electronic communication enlarged the possible scale of social organization and greatly enhanced the possibilities of centralization and imperialism in matters of culture and politics'.

Innis argued against technological determinism, and that to achieve a technology's potential requires sympathetic institutions and a culture that could effectively support alternative styles of life. He thus recommended the maintenance of alternative traditions of life and thought, traditions attuned to the demands of history and the need for stability and continuity. Carey and Quirk concluded that the task of intellectuals was to address themselves directly to questions of control of technology: not of centralism versus decentralism but of democratization and 'the reconciliation of immense power and wealth with the ideals of liberty and equality'.

However, the earlier enthusiasts for large-scale electricity projects failed to notice that large dams and power stations would be controlled by electric utilities, firms or government agencies destined to have enormous power. The anti-nuclear movement would take up this issue of the control of technology, with critics such as Patterson and Lovins challenging the autocratic decision-making of big centralized utilities, such as the CEGB. Instead they proposed small decentralized energy sources, based on solar energy, with the thesis that it was decentralization of generation that was crucial to democracy and community, not decentralization of use.

Techno-arcadia

Electrification was promoted in the inter-war years not only as progress for urban areas, but also as the saviour of the countryside, then suffering from a prolonged agricultural depression. This new form of energy it was argued could revitalize agriculture and rural life, and could be the saviour of the English tradition and way of life. As Luckin remarked, triumphalists came to stress 'natural' connections between farming, the revival of the 'organic' village community and the new form of energy as a stimulant to rural crafts and industries.

This dream of a 'techno-arcadia', long-standing in English and American utopian thought, would allow the decentralization of society and a dynamic, participatory and humanized economy of villages in which craft would be revolutionized and made more widely available. In this arcadia, the ravages of industrialism and the machine were to be ameliorated in a clean and communal agrarian environment, itself renovated and revitalized by electricity. In return it was hoped that the countryside would learn to love the high-voltage electricity pylons straddling the countryside, and that those who could not share the modernist's aesthetic tastes for them could at least see them as an essential part of rural life.

This dream was supported in the 1920s by Lewis Mumford who saw the potential for decentralized cities through the use of hydroelectricity. He later warmly praised, in his *The City in History* (1961), Kropotkin's early vision 'that the flexibility and adaptability of electric communication and electric power… had laid the foundations for a more decentralized urban development in small units, responsive to direct human contact and enjoying both urban and rural advantages'. Mumford also supported the 'village industry' concept of Henry Ford, the car manufacturer, who publicly stated that they could be as efficient and profitable as large-scale urban centres (and privately believed that they were less conducive to labour unions and strikes). While the countryside welcomed the benefits of electrification in the first half of the twentieth century, there were objections by conservationists on aesthetic grounds: the damage to the landscape from pylons and dams.

Opposition to pylons

The first opposition in Britain was in urban areas where city councils had opposed the proliferation of local distribution poles since the 1890s and wanted the lines placed underground. But there was not public opposition until the proposed construction of the National Grid from 1926, which involved building 4,000 miles of high-voltage pylons across rural areas. As Luckin remarked, the natural landscape and 'amenity', farming land and property rights – all would be threatened by this 'march of the pylons' and the building of massive and polluting 'super-stations'. Conservationists were quick to realize the dangers of these 'new and alien physical structures', and the countryside defender, Clough Williams-Ellis, wrote in 1928 in his book *England and the Octopus*: 'Clearly the electrification of England will not be accomplished without severe shocks to amenity… What is this magic network going to cost us?'

The triumphalists realized that this plan would attract hostility from rural dwellers affected by but not yet benefiting from electricity supply. However they were certain that the long-term economic benefits of the National Grid would outweigh the social costs, and that the 'rural community' must be willing to pay high short-term costs to achieve so 'majestic' a scientific and cultural transformation. Rural conservationists were not convinced however, and protest groups established themselves in the late 1920s and early 1930s in the South Downs, the Lake District, the New Forest and London. The strong opposition in Sussex led to a two-day public inquiry in September 1929 with much national media attention and debate, and lobbying of ministers. The National Grid plan was approved, which led to strong feelings of bitterness in rural Sussex at the lack of government attention to local opinion. The approval was seen as an unqualified defeat for the anti-pylon campaign, and led to congratulatory articles in the electrical press on how powerful landowners and articulate intellectuals had been outmanoeuvred. Subsequently from the early 1930s attitudes to pylons among conservationists and the 'aesthetes' began to change, towards greater acceptance that pylons were part of 'man-made nature'.

Opposition to dams

Britain has limited hydroelectric capacity, except in Scotland where the first scheme was completed in 1896 by the British Aluminium Company at Foyers near Loch Ness. But progress at developing Scotland's hydro potential was slow in the inter-war years, dogged by protests at private ownership from Scottish nationalists and at its aesthetic and landscape effects from conservationists. It was only with the creation of the North Scotland Hydro-Electric Board in 1943, dedicated to providing social benefits and economic regeneration of the Highlands, that there was substantial dam building.

During the 1930s there was great publicity for dam projects with dam engineers obtaining a heroic status. As Basalla commented, hydroelectricity was 'seen as a perfect, almost magical form of energy' which would 'lead to the decentralization of society and to the establishment of a utopian society'. This euphoria reached its peak during the New Deal era in the United States with the TVA intended to serve as a showcase for the positive linkage of electricity, decentralization and citizen participation in reclamation of the landscape. Naturally it failed to live up to its utopian expectations, and to Carey and Quirk it demonstrated the 'folly of identifying technical projects with the creation of democratic community'.

Post-war the TVA was promoted to other countries as a model of dam building for social democracy under the rhetoric of 'universal electrification', and more discretely as a weapon in the cold war – a role later played by nuclear power under the Atoms for Peace programme. In the post-colonial era of the 1950s leaders of the newly independent nations in Africa saw dams as symbols of modernity and independence, the most famous case being President Nasser and the Aswan dam in Egypt. Dams served political purposes, and as John McNeill, the environmental historian, remarked in his book *Something New Under the Sun* (2000):

> Communists, democrats, colonialists and anticolonialists all saw some appeal in big dams. Governments like the image they suggested: an energetic, determined state capable of taming rivers for the social good. Dams helped to legitimate governments and popularize leaders, something the United States needed more than ever in the depression years, and something Stalin, Nehru, Nasser, Nkrumah, and others all sought.

However there was increasing opposition by conservationists in Germany and the United States to the building of massive dams that would alter landscapes and adversely affect downstream ecosystems. The building of large dams became a focal point of national environmental protest because they were highly visible and disruptive projects. In Germany there were vocal campaigns before the First World War with renewed opposition only in the early 1950s. According to Raymond Dominick in his book *The Environmental Movement in Germany* (1992), there was a campaign against the building of dams in the Alps from 1908 until 1917, and then anti-hydro protest from 1947 to 1955 against the utility BAWAG, which had the slogan 'Without electricity no economic life'.

In the USA, the Sierra Club's successful campaign from 1950 to 1955 against the building of a dam at Echo Park in Dinosaur National Monument (on the Colorado-Utah border) marked a return by that organization to national campaigning after an absence of over three decades following its damaging and divisive campaign against the Hetch Hetchy dam. The Hetch Hetchy dam controversy has assumed mythic status amongst many commentators, a battle between the 'utilitarians' – those wanting to develop resources scientifically – and the 'conservationists'– those wanting to protect national parks from any development. The battle was a key mobilization point for conservationists around the country, but the Sierra Club, despite John Muir's efforts, was split on the issue, which involved for Californians many complex political and social issues over water rights, the public control of utilities, and the role of national parks. Muir, who died in 1914, soon after the battle was lost, was seen by many as a martyr for the conservationists' cause, and a victory did come with the creation of the National Parks Service in 1916. However the Sierra Club did not fare well, there being a significant minority opposed to Muir's position – a Sierra Club poll of its members in 1910 voted 589 to 161 for Muir's position – and eventually about 50 members resigned. After the death of Muir the Sierra Club became little more than a regional hiking club, and was only rejuvenated in the 1950s with the appointment of David Brower as its first executive director. Together with the Wilderness Society, Brower led a five-year campaign to prevent the building of Echo Park dam.

The significance of the conservationists' victory at Echo Park was public awareness for the first time that the national parks and monuments had been created for the use and enjoyment of all people and were thus inviolate. From then on most conservationists opposed dam building and instead supported nuclear power. As historian John Wills commented in his paper 'Welcome to the Atomic Park' in the journal *Environment and History* (November 2001):

> The American conservation lobby vilified hydroelectric projects as concrete behemoths threatening large-scale disruptions of river ecosystems, while welcoming talk of ecologically benign, self-contained atomic energy facilities. However support for the peaceful atom wavered when atomic developers chose sites of specific interest to the conservation lobby.

In the early 1960s the siting of nuclear power stations and their potential impact on scenic resources caused controversy for the Sierra Club. The first site proposed by the Pacific Gas and Electric (PG&E) utility on the Californian coast was at Bodega Bay, which was eventually abandoned as unsuitable after the discovery of the San Andreas Fault beneath it; the site was turned into a nature reserve. Some of the Sierra Club's executive then worked closely with PG&E to find an acceptable site, eventually deciding to support PG&E's choice of the Nipomo Dunes. David Brower and others protested at the club's endorsement of Nipomo Dunes, questioned its decision-making process and whether the club should participate in such deals. A bitter organizational conflict ensued which almost split the club, and led to the firing of Brower and a shakeout of the club's leaders.[7] Nipomo Dunes was abandoned in favour of the site of Diablo Canyon, where a reactor was built.

There was similar opposition in Britain to the use of remote coastal sites for nuclear power stations, particular by Max Nicholson at the Dungeness inquiry in 1958. The post-war building of hydroelectric plants in Wales was opposed by conservation groups such as the CPRE on amenity and landscape grounds, particularly over developments in Snowdonia National Park. However by 1971 the CEGB, through its attention to environmental impacts, encountered much less opposition with its next large pumped storage scheme at Dinorwig.

Solar energy: a long-standing dream

The sun and its life-giving powers have long-standing religious and mystical associations dating back millennia and the belief in the possibility of a utopia powered by solar energy dates back to the Renaissance if not beyond.[8] This belief faded in the early twentieth century, as the appeal of hydro and nuclear power asserted themselves, but was reborn in the late twentieth century as these two energy sources fell from favour. Solar energy, renamed 'natural', 'renewable' or even 'sustainable' energy, is now seen as an essential part of the move to a 'sustainable' society which while not utopian still possesses much ecological and social virtue. Solar energy is used in its widest sense, encompassing energy not only directly from the sun but also indirectly from wind, waves, the tides and water (but not from large dams), as well as from biomass sources (such as from trees or agricultural residues).[9]

Nineteenth century solar utopias

Given that until the Industrial Revolution, and the widespread use of coal, there was no life, heat or power without the sun, it is entirely understandable that old utopias were centred around the sun, and that the ancient myths and symbolism of the sun as the source of life, wealth and health should persist until today. Intimately associated with solar energy is the idea of small communities, generally agricultural and decentralized, living in 'harmony with nature', and this was a staple of utopian fiction in the nineteenth century.

One utopian work that gave detailed consideration to using solar energy was by John Aldolphus Etzler, a German living in Pittsburgh. In *A Paradise Within the Reach of All Men, Without Labour, by Powers of Nature and Machinery* (1833), Etzler proposed the creation of a paradise based on material luxury within ten years simply by harnessing the planet's cost-free energy sources – the wind, the waves, the tides and the sun. His pamphlet contains detailed mathematical calculations of the power that would accrue and simple descriptions of the machines needed to access and store it. His belief was that 'Nature affords infinite powers and wealth', and its exploitation should be the basis for a more perfect society. However as John Carey, in his book *The Faber Book of Utopias*, commented: 'Unlike modern advocates of natural energy, Etzler was not concerned to preserve the environment. Very much the contrary: he wanted to change it entirely. Mountains will be flattened, rivers will flow in vitrified channels, deserts will flower'.

Etzler proposed that the natural world would be completely remoulded to man's taste: that America's large forests be 'ground to dust' and then 'cemented by liquor' to provide a 'universal building material' rather like plastic wood. This would then be moulded by machinery to any shape and, if required, vitrified, so that it would be virtually indestructible and would 'radiate with a crystal-like brilliancy'. Mankind would live in an immense sculpture park, the greenery tastefully interspersed with porticoes, foundations and statues, all dazzling to look at. Dolores Greenberg, an American historian, believed that Etzler's treatise had a big influence on later nineteenth-century utopian thought, particularly on Robert Owen, especially Etzler's idea of 'abundance without labor, the product of harnessing the energy of the sun, the tides, the water and the wind'. However Etzler's idea of utopia as luxurious living did not appeal to all, especially Henry Thoreau. He read it with 'deep disfavour' and in an 1843 review deplored its emphasis on 'gross comfort and pleasure at the expense of the higher life'. According to Carey it seems to have set Thoreau thinking, and little more than a year later he built his wooden hut on the edge of Walden Pond and began his two-year retreat from materialism and the pressures of civilization, described in his classic *Walden, or Life in the Woods* (1854).

By the late nineteenth and early twentieth century, due to technical advances and mass production, renewable energy devices had become widespread, particularly windmills for water pumping in the North American plains, solar water heaters in California and Florida, and solar power plants to produce steam. There was also electricity production from hydro and geothermal power. The 'Back to the Land' movement in Britain in the late nineteenth century praised the virtues of small communities and the preservation of nature, and saw electricity from hydropower allowing an escape from cities to the countryside. The outlook for solar energy looked bright, and writers chose it for their rural utopias, contrasting it to the evils of urban societies based on coal. William Morris, in his famous utopia, *News from Nowhere* (1890), has society powered by electricity generated from wind and water power, while Jules Verne in his 1890s short story, *In the 29th Century*, uses wind, water, solar and geothermal energy, and H. G. Wells has wind and water power (monopolized by the Windvane and Waterfall Trust) in *When the Sleeper Awakes* (1899). In *The Case of The Fox* (1903) William Stanley had his utopia powered by a multitude of renewable sources of energy.[10] In countries where coal had to be imported, such as Egypt, solar power plants received much encouragement from the colonial governments. The solar pioneer and visionary, Frank Shuman, was by 1914 confident that his optimistic predictions were going to be realized and he wrote (quoted in *A Golden Thread* by Butti and Perlin in 1980):

> Sun power is now a fact and no longer in the 'beautiful possibility' stage... [It will have] a history something like aerial navigation. Up to twelve years ago it was a mere possibility and no practical man took it seriously. The Wrights made an 'actual record' flight and thereafter developments were more rapid. We have made an 'actual record' in sun power and we also hope for quick development.

However after the First World War interest in solar power quickly declined due to the discovery of new sources of oil in Iraq and Iran. There was soon a world oil and gas glut, and solar energy disappeared from the public view until after the Second World War.

Solar versus atomic energy

After the Second World War there were fuel shortages and rationing in the USA and Britain, particularly for domestic consumers. There was a great debate about energy for the future, with warnings on oil running out, the possibility of cheap atomic power and the promise of solar energy. This debate was also set against public concern, raised by conservationists in such books as *The Road to Survival* (1948) by William Vogt and *Our Plundered Planet* (1948) by Fairfield Osborn, about the dangers of world population growth and limits to food and natural resources. Thus the warnings in 1949 by King Hubbert, an oil geologist, on future fossil fuel shortages attracted much attention and Palmer Putnam was commissioned in 1949 by the AEC to write a report on long-term (50 years) energy prospects. Putnam, whose book *Power from the Wind* (1948) had been published the previous year, had much personal support for solar and wind power, but in his report he judged solar energy as too expensive, and the future energy solution he believed was with the development of nuclear power and the electrification of the economy. This view was opposed by Farrington Daniels, a keen advocate of solar energy and a leading contributor to the debate about the merits of atomic and solar energy. In an article 'Our Energy Resources' published in *Physics Today* in April 1949, Daniels and his colleagues agreed with the conservationists, writing that 'humanity is on an unrepeatable spending spree, that we are using up minerals and energy resources on an unprecedented scale, and that these will run out in the foreseeable future'.

However the solar dream faded with the easing of post-war fuel shortages. New oil and gas resources were developed, and the US government committed itself to nuclear power. Nevertheless solar advocates persisted, forming the Association for Applied Solar Energy in 1955, followed up by a World Symposium in Phoenix, Arizona in 1956. Until 1973 this was to be an era of cheap oil, but despite this there was a continuous stream of books promoting solar energy, such as *Power from the Sun* (1957) by D. S. Halacy, *Direct Use of the Sun's Energy* (1964) by Farrington Daniels and *Solar Energy for Man* (1972) by Brian Brinkworth. Solar energy did not however disappear from policy view, but it was just considered as a 'backstop' or 'insurance' energy source in case of fossil fuel shortages or the failure of nuclear power. Typical of this viewpoint is Dennis Gabor who wrote in *Inventing the Future* (1963):

> There is no need, however, to base our hope for abundant energy supplies on fusion power only. We have seen that the free and inexhaustible sources of power – wind, water, tides, geothermal and solar energy – are either too small or too unprofitable at the present time. But they could be large enough for a reasonable humanity, which prefers civilization to numbers, nor will they be unprofitable at a time when oil is no longer there to gush out of the earth. Civilization need never be in danger for lack of power.

Solar as hip

Until the late twentieth century energy utopianism was built on the idea of harnessing the power of nature for the benefit of mankind, without regard to the environmental consequences. There were of course critics of these adverse consequences, from campaigners against coal smoke in the nineteenth century to opponents of dam building in scenic mountain areas in the early twentieth. But few, if any, questioned the need for new sources of energy. Higher energy consumption was associated with an increased standard of living, a better quality of life and with 'progress' in general (as it still is in most developing countries keen for rural electrification).

The nuclear dream was the most explicit of these utopias, based on (masculine) mastery of (feminine) nature, an endless supply of cheap energy, and luxurious living. This was a continuation of Etzler's vision a century before, though now the outdoors was turned into a wilderness park for off-road vehicles rather than an aristocrat's country park. Whether energy came from nuclear or solar power made no difference to the attitudes of society towards the environment: it was there to be exploited and moulded to man's tastes.

What, however, was new and different about protest against the nuclear dream from the late 1960s was the attitude to consumption. Less was now to be better than more: small was to be beautiful. Solar energy was now a means to 'drop-out' from the materialism of society, and to win independence from 'the system'. The symbolism of solar energy was immense: it was a tool for survival, a means of subsistence, and an act of rebellion. The old environmental texts by Victorian writers such as Kropotkin, Morris and Thoreau were rediscovered and mixed in with more recent ones by Mumford, Huxley and Schumacher. The utopia was now to be based on decentralized, small-scale communities using 'appropriate' technology with a reverence for nature. A new world could now be created based on (small-scale) solar energy. What such an 'ecotopia' would look like became the stuff of 1970s science fiction writers such as Doris Lessing and Ernest Callenbach, and the future of a post-apocalyptic world now became sunny.

Public interest in solar energy as a utopian energy source was started by the counterculture in the late 1960s in the USA. One of the first examples was at Drop City, a drop-out artist community in Colorado where, inspired by the hi-tech utopian writings of Buckminster Fuller, they built domes and a solar water heater. This idea of using free energy and DIY technology received wide publicity, and solar energy received extensive mention in Stewart Brand's seminal publication *The Whole Earth Catalog*, first published in 1968. Brand set out to create a survival manual, and extolled alternative energy and the virtues of disconnecting yourself from the grid. In the *Catalog* Steve Baer[11] reviewed Farrington Daniels' book, *Direct Use of the Sun's Energy*, and said that it was: 'The best book on Solar Energy that I know of... Any curious and intelligent person can learn a great deal about our planet and ourselves by reading this book about ways of using sunlight... I read the book on a Greyhound bus in Texas two years ago and it has changed my life and my way of thinking'.

Solar energy was now not just for physicists but could be used by anyone with simple technical skills seeking energy independence. Technology, not rioting on the streets, was now the key to 'revolution' in society. As Langdon Winner commented the obsession with technology arose about the same time – from 1969 – as a general disillusionment with protest politics that had become too dangerous or depressing. He wrote: 'It was during this period that many in the United States dropped out of political activity and began a certain kind of sociotechnical tinkering: roof gardens, solar collectors, and windmills became a focus of community action'.

It was believed that what was needed was a new type of technology, one that if used morally and ecologically could create a utopian society. This technology went by a wide variety of names, the earliest of which was 'Intermediate Technology' coined by Fritz Schumacher in 1964. For the next ten years a group of countercultural radicals, environmentalists, scientists and social activists discussed the concept and invented names such as appropriate, alternative or radical technology, often shorted to 'AT'. There were numerous definitions of AT, each with long lists of attributes, but basically all agreed that AT had to be cheap, simple and ecologically safe. Above all it had to use solar energy.

There was an immense amount written on AT in the 1970s mostly by enthusiasts, with recollections by two participants (Witold Rybczynski in his book *Paper Heroes* and Langdon Winner in his article 'Building a Better Mousetrap') both appearing in 1980.[12] AT merged into the 'soft energy' movement based on energy efficiency and renewable energy, popularized by Amory Lovins. This initially had a strong belief, like AT, that the choice of energy technology could radically alter society, but its utopian sentiments had to be toned down in the 1980s in order to succeed under the 'free market' ideology of the Reagan and Thatcher governments.

Since then utopian sentiments have been strongly suspect, but solar energy now respectable as renewable energy is still being promoted as the solution to many of society's problems. In one of the few retrospective analyses of AT, the environmental historian Andrew Kirk remarked that the AT movement's thoughtful re-evaluation of the role of technology in society 'is perhaps the most significant and lasting contribution of the counterculture to American culture'. Despite the high hopes of solar enthusiasts in the twenty-first century, couched in the new utopian language of global sustainability, Basalla's words of warning against believing in energy utopias may still be relevant. Solar energy myths, he warned, 'have been exposed as naive and unrealistically optimistic. The energy of the sun will be utilized, but it will not be done soon; it will be more costly than we can imagine; it will not be without environmental problems; and it will not form the basis for a new utopian way of life'.

Chapter 6

Ecology and science fiction

In this chapter I would like to pose the question, 'Is science fiction the literature of the environmental movement just as nature writing was the literature of the conservation movement?' Comment on – and sometimes solutions to – society's problems are the province of fiction, and this chapter explores the impact of a whole genre of popular literature, science fiction (SF). This area of fiction was very popular with young people post-war and has been ignored by most writers on environmental movements. This is in contrast to the widely accepted view that 'ecological' literature or 'nature' writing has had a formative influence on the development of environmentalism, and vice versa. This neglect of SF contrasts with the view that SF has been a powerful force in shaping societal attitudes to the atomic bomb, given prominence by such writers as Paul Boyer and Spencer Weart.

Only a small proportion of SF gave any consideration to environmental problems or solutions, but the few novels that did often sold in the millions. Their importance is in disseminating a favourable image of ecological concepts of harmony with nature, and of population control. Frequently they were published at the same time as less popular works by conservationists, and in the SF novels there is often reference to post-war conservation literature. Thus I believe SF novels reinforced the conservation message and created favourable impressions amongst many young readers, which could then be used to recruit them to the environmental movement in the late 1960s and early 1970s. The message left in young minds was that 'ecology' and 'conservation' should be important components of life in the future. At its most powerful SF can instil visions of the future into young impressionable minds that last a lifetime.

SF as literature

The role that nature writing played in the period between 1800 and the 1950s in promoting the protection, preservation and conservation of wildlife and nature has been widely acknowledged. Greatest, and perhaps the last of, such writers was Rachel Carson, whose classic *Silent Spring* (1962) bears witness to the destruction of nature she so passionately loved. However the readership of such books was mainly middle-aged and middle-class countryside lovers, whose fears about over-population were frequently dismissed as 'reactionary' and 'elitist': more concerned with the plight of animals than humans. However in the 1960s there were new developments, firstly the 'New Wave' in SF which explored the impact of technology upon people, and secondly the 'ecologists' who were concerned about the impact of technology upon nature and society. An important factor also, particularly in the USA, was the increasing academic respectability of SF literature, now a suitable subject not only for PhDs but also for secondary school reading lists. It became part of the school curriculum basically because teachers found that children would only voluntarily read this type of literature. Thus SF (and fantasy literature) was very popular amongst young people and became an integral part of their counterculture particularly in rock music and in the underground press. This youth was to form the mass of the emerging environmental movement in the early 1970s – and SF increasingly reflected their ecological concerns.

However in analysing SF literature it must be remembered that SF is part of commercial publishing, it exists to be sold; hence sales are more important than literary quality. As Harry Harrison, a noted SF author, pointed out in his chapter in *Future Imperfect*, 'It has to be entertaining because it's the thinking man's garbage, the thinking man's entertainment'. Like all literary genres that aim to sell to a mass market, such as 'westerns' or 'romance', there is a great deal of rubbish. As SF historian Chris Morgan commented in his book *The Shape of Futures Past*, it is 'inevitable that a great many tales of the future should be substandard, partly because there is a low-quality end to every genre of fiction and particularly because science fiction has been a rapidly expanding market area with a largely undiscriminating readership'. SF sells in the tens of millions and has attracted a wide range of authors, from mid-nineteenth-century nature writers to late twentieth century feminists. Thus it is no surprise that there are some SF books of real literary merit on ecological themes, which will be discussed later in this chapter.[13]

The impact of SF on society and writers

As a genre SF has attracted many authors who have seen it as an ideal medium for political satire and criticism, and it is still the natural choice for the dissemination of utopian and dystopian ideas.[14] Some novels have been very influential in their impact on society, shaping the political discourse about the nature of society and the role of technology. Perhaps the two most important novels in terms of effect were Edward Bellamy's *Looking Backward 2000-1887* (1888) and Theodor Hertzka's *Freeland* (1891).[15] These were both very influential, selling large numbers of copies in many countries and leading to the foundation of societies which tried to put their respective philosophies into practice. Bellamy's book was the second bestseller of the nineteenth century in the United States, after *Uncle Tom's Cabin*, selling well in excess of 250,000 copies in the USA alone. In Britain by October 1891 – two years after publication here – it was already into its 22nd edition and was translated into many European languages. The book led to an intense debate on the nature of the utopian socialist society and in Krishnan Kumar's opinion influenced many political and scientific writers, such as Alfred Wallace, Ebenezer Howard, William Morris – who in response wrote *News from Nowhere* (1891) – and H. G. Wells.

Kumar believed that it is difficult to gauge the impact on particular individuals, but *Looking Backward*, he argued, had an extraordinarily widespread appeal for a particular generation of intellectuals and created a vogue for utopias, with sixty-two being written in the years 1889-1900. Bellamy's idea of an all-encompassing 'communist state', with a high standard of living based on the use of technology, rejected previous ideas of communitarian socialism based on small self-sufficient communes. It was also to encourage the concept amongst intellectuals of large centralized energy sources such as hydropower (and later nuclear power), rather than small-scale decentralized solar energy. Henceforth the socialist state was to be based on large-scale electrification of society, or in Lenin's famous slogan 'Electricity plus soviets equals communism'. The 1920s and 1930s were years of ceaseless missionary activity by the 'science and society' movement, which chimed in with a wider political commitment to centralized planning on a national scale. As Kumar remarked, 'the scientists moreover, left no one in doubt that their claims were not merely on behalf of a superior technique or method but implied a new politics'. Through unending proselytism such scientists ensured that to most intellectuals science and socialism were intimately connected, almost synonymous.

This 'scientism' was reflected in the wildly technologically optimistic tone of most SF magazines of the time. Hugo Gernsback, the editor of the US magazine

Amazing Stories, was its chief propagandist, and as Brian Stableford wrote in the 1979 edition of *The Encyclopedia of Science Fiction*:

> He called this medium 'scientifiction' which he saw as a means of anticipating the transformation which the world was already undergoing through the acceleration of technological progress... In the editorials, which he wrote for his early SF pulps, he talked about atomic energy and radar and television and space travel... The effect of SF upon young readers in the 1920s and `30s – which was tantamount to a revelation or quasi-religious conversion in many cases – was mainly due to its opening up a consciousness of the *immediacy* of change.

The impact of this quasi-religious conversion is very difficult to explore, as few intellectuals (except perhaps in old age) are prepared to admit that their adult views were shaped by their childhood reading, particularly of a genre that had a reputation as trash. It can only be surmised on anecdotal evidence and I can only put forward two observations[16] (the first of which got me interested in this topic in 2000):

A newspaper interview in *The Independent* of 9 June 2000, with John Maynard Smith, the evolutionary biologist, which started, 'In 1931 a science fiction writer called Olaf Stapledon wrote a book which changed the lives of two men, *Last and First Men*'. The two men were Arthur C. Clarke and John Maynard Smith. The interview claimed: 'Were it not for Stapledon's book, Maynard Smith might have gone on to become a stockbroker... Instead he became a communist'.

John Davoll, the Director of the Conservation Society – Britain's first environmental society – in an interview with me in 1999 said that as a boy Stapledon had a 'big impact' on him and that he joined the Olaf Stapledon Society.[17]

Olaf Stapledon is a long-neglected philosophical writer who earned a huge (and completely unexpected) following and popularity amongst SF readers. His first work of fiction, *Last and First Men* (1930), caused something of a sensation and was widely acclaimed by contemporary writers and critics, but has been forgotten by literary historians. His work still has a cult following and there are several websites devoted to it.[18]

The link between ecology and SF

The development of ecological concerns by scientists was quickly reflected in the SF literature, albeit in a sensational way. Emerging ecological concepts such as the interaction of species and the dangers of insecticides, pollution, over-population and resource shortages were taken up rapidly by SF writers, usually in the form of 'eco-catastrophe' plots. Generally these had little or no ecological discussion and basically the eco-catastrophe was a device to write a 'post-apocalypse' novel. However there were a few writers who used it to implicitly warn readers of current ecological dangers and suggest solutions. To them the future was dystopian unless action was taken. Others however were more utopian, using this plot in order to rid the world of modern industrialized society and return to a more pastoral and (to them) more desirable society.

Table 1 shows a chronology of ecological and SF literature, and it is interesting to note that several authors, including Paul Ehrlich and Garrett Hardin, wrote both types, generally the SF novel to publicize their ecological views. Charles Elton, in 1927, was the first writer to publicize the word 'ecology' and its concepts, particularly on the complexity and interrelationship of life. Two years later, in 1929, came what has been described as the first ecological SF story, *The Man Who Hated Flies* by J. D. Beresford.[19] This is a parable of a man who discovers a perfect insecticide and precipitates an ecocatastrophe by removing pollinating insects.

In the late 1930s and 1940s, there was a spate of ecological books dwelling on the world problems of over-population, food shortages and soil erosion. Agricultural writers such as Stapledon, Lymington and Jacks and White produced sombre and serious books, which were followed by more populist and sensational works from American conservationists such as Fairfield Osborn and William Vogt. Overall these books had a distinctly Malthusian message and were deeply influenced by the global agricultural depression and the dust bowl era in the 1930s. These Malthusian visions, combined with the dystopian visions of the late 1940s of totalitarian regimes and nuclear destruction, produced two contrasting (and bestselling) SF novels in the early 1950s, *Earth Abides* and *The Space Merchants*. The former welcomes the end of (American) industrial society and the return to nature, whilst the latter pits the 'conservationist' radicals against big business (in the form of advertisers) in an over-populated world for control of the future. Kornbluth followed up these themes in 1958 with *Shark Ship*, a horror story of an over-crowded and polluted world. Similarly J.G. Ballard in his short story *Billenium* (1961) dealt with the social consequences of global over-population for city life, a theme used by Harry Harrison in his bestseller *Make Room! Make Room!* (1966). However the usual post-war reaction from SF writers was to use the 'eco-doom' messages as a new scenario for the 'eco-catastrophe' novel, with by far the greater emphasis on dystopian image of the future than on ecological issues.

Early examples include *Greener Than You Think* (1947) by Ward Moore and *The Death of Grass* (1956) by John Christopher. In both a blight affecting grass species precipitates world catastrophe (though this theme is a reworking of the 1923 SF novel *Nordenholt's Million* by J. J. Connington). Other writers who used the 'eco-catastrophe' scenario – whereby humans suffer due to the misuse of nature and technology – were John Wyndham (e.g. *The Day of the Triffids* (1951)), and J. G. Ballard (in such novels as *The Wind from Nowhere* (1962) and *The Drought* (1964)).

Table 1 Selected Chronology of Ecological and SF Literature

Ecology	Science Fiction
1927 Elton, *Animal Ecology*	
	1929 Beresford, *The Man Who Hated Flies*
1935 Stapledon, *The Land: Now and Tomorrow*	
1938 Lymington, *Famine in England*	1938 Street, *Already Walks Tomorrow*
1939 Jacks and Whyte, *The Rape of the Earth*	
1948 Osborn, *Our Plundered Planet*	
1948 Vogt, *The Road to Survival*	
	1949 Stewart, *Earth Abides*
	1953 Pohl and Kornbluth, *The Space Merchants*
1954 Osborn, *The Limits of the Earth*	1958 Kornbluth, *Shark Ship*
	1961 Ballard, *Billennium*
1962 Carson, *Silent Spring*	
1964 Hardin, *Population, Evolution and Birth Control*	
	1966 Herbert, *The Green Brain*
1966 Commoner, *Science and Survival*	1966 Harrison, *Make Room! Make Room!*
1968 Ehrlich, *Population Bomb*	1968 Brunner, *Stand on Zanzibar*
	1969 Ehrlich, *Ecocatastrophe*
1970 Rattray-Taylor, *The Doomsday Book*	
	1971 Sauer, ed. *Voyages...*
	1971 Disch, ed, *Ruins of the Earth*
	1972 Brunner, *The Sheep Look Up* Wylie, *The End of the Dream*
	1972 Hardin, *The Voyage of the Spaceship Beagle*
	1973 Elwood and Kidd, ed. *The Wounded Planet*
1974 Kropotkin, *Fields Factories...*	1974 Le Guin, *The Dispossessed*

Return to nature

The idea of a 'return to nature' has a long tradition in British culture and was used by two Victorian nature writers and distinguished naturalists, Richard Jefferies and W. H. Hudson, who turned their literary talents to SF in the 1880s.[20] As I. F. Clarke commented:

> Jefferies and Hudson had much in common: they were both distinguished natural-
> ists; they had evolved new styles of writing about life in the country... they shared
> a most vigorous detestation of the squalid, dreary, unsatisfying life that had
> become the norm for the masses in the great new towns. In violent and agreeable
> anticipative acts of destruction the two men called on nature to wipe out the infamy
> of urban civilization.

Richard Jefferies was well-known for his eulogies of the countryside, and in his novel *After London* (1885) he delights in describing most of the country (but espe-cially London) being reclaimed by nature as the few people who remain relapse into barbarism, or in the words of Kumar, 'a new Dark Age, a world of poverty and cruelty, of warring petty tyrants and marauding robber-bands'. While the novel is clearly anti-utopian, Jefferies is ambiguous – for while he hated London and all that it stood for and thus longed for the 'Great Purge' of industrial civi-lization he believed that a return to rural poverty and violence was no answer.

More positive was W. H. Hudson in his novel *A Crystal Age* (1887) which created the prototype of the modern 'Arcadia' of the future.[21] The novel rejects almost every aspect the Victorian age, and imagines a sparsely populated world at least 10,000 years after a catastrophe has almost wiped out mankind. In this ascetic and matriarchal pastoral society, small groups of people live a close family life in a Rousseau's paradise of tame animals, handicrafts and the cheerful culti-vation of their gardens. It was not a popular book but its innovation was to introduce the 'apocalyptic Arcadia' to the tale of the future, and to reintroduce the ancient myths of man in harmony with nature that first appeared in Greek poetry, myths that would then reappear in SF over fifty years later.

Victorian visions of the city

The most famous and widely read arcadian utopia of the late nineteenth century was *News from Nowhere* (1890) by William Morris, who had been deeply affected by the pessimistic portrayal of medieval life in Jefferies' novel *After London, or Wild England*.[22] In Morris' novel, which he wrote as a response to Bellamy's *Looking Backward*, everything is small-scale, local and anti-technological. In the view of Chris Morgan his society is 'patently unworkable for any community larger than a village... but this was just the sort of society which Morris was ideal-izing, in contrast to Bellamy's soulless and hive-like recommendation'.

Although Morris idealized medieval village life, most social reformers and utopians were concerned with the city of the future, and above all with ending squalor, pollution and dirt. The city of the future, in virtually all utopian visions, was remarkable for its cleanliness and open spaces, and in this they were all influ-enced by the work of Dr. B. W. Richardson, the leading British expert on public hygiene. His pamphlet *Hygeia: or, A City of Health* (1876) was widely read, and his ideas were adopted by, among others, Edward Bellamy and Ebenezer Howard. It is also probable that Ebenezer Howard in addition read Bellamy's *Looking Backward* and perhaps Hertzka's *Freeland*, before deciding that utopia need not be built on the site of an old city but on greenfield sites 15-20 miles from the centre of London. Howard's seminal work in the field of town planning, *Tomorrow: a Peaceful Path to Real Reform* was first published in 1898, and an amended version issued under its more famous title *Garden Cities of Tomorrow* in 1902.

While social reformers and utopians in the late nineteenth century were seeking the rebirth of the city, SF writers were imagining its end, a death brought about by its waste products. One of the first SF stories about a catastrophe brought about by pollution was *The Doom of London* (1892) by Robert Barr where a smog suffocates the city. However this is in the genre of the 'eco-catastrophe' novel where an ecological disaster is a convenient device to use in a 'post-apoca-lypse' novel. SF novels from authors such as Jefferies and Barr reflected late nineteenth century concerns that industrial civilization was doomed by its noxious wastes but may also be seen as part of the new wave of 'the literature of terror' inspired by the 1890s Gothic imagination. Such novels as *Dr Jekyll and Mr. Hyde* (1886), *The Picture of Dorian Gray* (1891), *The Island of Dr Moreau* (1896) and *Dracula* (1897) instilled in the popular imagination major myths and archetypes of hideous and invincible powers, which sprang from the very nature of man and the cosmos, and to which science and civilization were no defence.

Fantasy literature

These tales of what we might call the supernatural opened the way to what SF critic Patrick Parrinder calls in his book *Science Fiction* 'an anti-scientific fantasy literature using the settings and situations of SF, but resorting to magic and sorcery to resolve any difficulty'. Fantasy (or 'sword and sorcery') literature is considered a separate genre from SF, but there is no doubt about its impact on some modern SF writers (such as Ursula Le Guin) and the modern environ-mental movement.[23] The most famous, and influential, of these writers were J. R. R. Tolkien and C. S. Lewis, both dons at Oxford in the 1930s. Tolkien's first novel, *The Hobbit* was published in 1937 and was uncontroversial, being considered a children's book. In contrast the 'cosmic' or 'space' trilogy from Lewis attracted intense criticism.[24] Lewis was an outspoken proselytizer for the Christian faith, which he believed was being undermined by secular values, particularly science. He deliberately used the genre of fantasy in order to publicize his beliefs, and in his dedication to *Out of the Silent Planet* acknowledged his debt to 'H. G. Wells' fantasies'. As Patrick Parrinder commented, 'Lewis's space trilogy... was intended as an attack on 'scientism' – the term he used to denote uncritical acceptance of scientific aims and methods as good in themselves'. The sense of the damage these books could do to the scientific cause led J. B. S. Haldane, one of the foremost exponents of the scientific world view, to attack the Lewis trilogy in an article in *Modern Quarterly* (1946) but Lewis did not respond in print. In *The Abolition of Man* (1943), Lewis had set out the philosophical position of his trilogy and his belief that a commitment to the subjugation of nature that is fundamental to modern science is necessarily a commitment to the subjugation of 'man' as well. Parrinder remarked: 'In answer to the technologists' programme of human "abolition", Lewis dreams wistfully of a "regenerate science" which resembles the most extreme yearnings of present-day ecologists'. Meredith Veldman observed that 'Lewis sought to highlight not only God's intended harmony between humanity and nature but also the sanctity of nature itself... Like many ecologists two decades later, Lewis asserted nature's right to exist, its holiness as a living entity, apart from its utility to humanity'. Due to problems at his publisher, Tolkien's trilogy *The Lord of the Rings* was not published until 1954-55, though Tolkien had written it in the period 1937-49. It was well received by critics on publication, but it was not until the mid-1960s that it achieved widespread popularity, blossomed into the major book cult of the twentieth century and has now become an outstandingly successful film trilogy.

Mid-twentieth-century visions

From the 1920s to the 1970s most SF was pessimistic, and this was a reflection of the then current fears of what the future might bring. It was an era of political dictatorships, world war, mass unemployment and rapid technical and social change. However this pessimism may also have been caused by the personal outlook of its readers. Gerald Klein, a critic of SF, believed that the near-universal distrust of science and scientists in SF of the 1970s was not due to the inherent nature of science itself, but due to the particular fate of the social group – the 'technologically orientated middle-class' – to whom most SF writers and readers belong. Klein's argument, made in *Science Fiction Studies* in March 1977, is that after going through periods of expansionist optimism (pre-1940) and of 'confident scepticism' (1940-60), the technological middle-class found itself increasingly in the position of a proletariat: essential to the working of modern industry, but further than ever from assuming the political and economic control of affairs which Wells and his disciples demanded. As Parrinder commented on Klein: 'Science, it is now clear, is not a revolutionary force in society, but a servant of big business and the international corporations'. This relationship of science to big business is an essential theme of much SF, explored in such novels as Vonnegut's *Player Piano* (1952), Young's *The Rise of the Meritocracy* (1958) and Le Guin's *The Dispossessed* (1974).

Since the end of the First World War the scientifically progressive utopia had become a rarity in SF, except for H. G. Wells' *The Shape of Things to Come* (1933), later made into an influential film, *Things to Come* (1936). The vision of a new society, arising after a holocaust, was mostly not a scientific utopia but a pastoral one. As Clarke wrote, 'Most utopias rejected the old compact between science and society. A new college of prophets revived the old arcadian survival myth of Hudson's *A Crystal Age* in their accounts of the simpler life that would follow the obliteration of industrial civilization'. Following the Second World War there was a further decline in technological optimism and disillusionment with the future amongst SF writers (apart from Mack Reynolds, and perhaps Arthur C. Clarke), caused by perceived problems over the rise of totalitarian regimes, the threat of the bomb and the dangers of over-population. Technology was now the problem, not the solution, and it became fashionable and then hackneyed to show, as Chris Morgan remarked, 'how the many tentacles of technology are oppressing Mankind'. SF writers turned inwards in their utopias, rejecting the animosities of world politics for the cultivation of close personal relationships and loyalty to the chosen community.

These were now small, often isolated island communities, which in the pastoral tradition of W. H. Hudson and William Morris sought to ban or strictly control all technology. The intention was not, as in the old utopian tradition, to put forward a feasible plan, which could be widely adopted, but to define the 'good life'. Thus utopian tales increasingly absorbed the moral purpose of the fantasy literature of Lewis and Tolkien, starting with *Seven Days in Crete* (1949) by Robert Graves, which is a cheerful tale of the 'good life' with contented peasants ruled by benevolent witches. Post-war SF showed a nostalgia for a rural way of life and this genre came to be known as 'pastoral' SF. Yearning for a past way of rural life is of course not confined to SF writers, but is a common theme in most western societies and is generally highly romanticized, except by rural writers such as Richard Jefferies. The anti-urban theme was common in much American SF that idolized small-town life of the horse-and-buggy days, particularly by

Clifford Simak who was described by the critic Kingsley Amis in his book *New Maps of Hell* (1963) as 'the science-fiction poet laureate of the countryside'. This glorification of a (now vanished) rural life is found in *The Waveries* by Fredric Brown (1945), *Dandelion Wine* (1957) by Ray Bradbury, *Alas, Babylon* (1959) by Pat Frank, and *Joyleg* (1962) by Ward Moore and Avram Davidson. As David Pringle wrote in the *Encyclopedia of Science Fiction*:

> What these works have in common is an emphasis on the virtues (and sometimes the constraints) of the rural way of life. They are, explicitly or implicitly, anti-city and anti-machine; they frequently extol the values of living close to Nature, of being in rhythm with the seasons. This bucolic and Luddite strain in genre SF has its origins in some major works of American literature such as Thoreau's Walden.

Pastoral SF declined in the 1960s with the rise of the 'New Wave' of SF writers who took a less lyrical and more hardheaded view of the future. Chief amongst these writers was J. G. Ballard whose visions of 'eco-catastrophe' were presented not in terms of ecology but Jungian psychology; his tales were about the personal experience of disaster and Patrick Parrinder describes him as the 'poet of the scientific world-view in decline'.

In *The Wind from Nowhere* (1962) Ballard has the theme of man's futile struggle against nature, with his survival instincts dulled by civilization and hubris. The plot is of the wind increasing slowly and steadily in speed until it destroys all life on earth with only a few survivors sheltering in underground military bases. Hardoon, a crazed industrialist and the villain of the book, has built a tower to defy the wind, and proudly declares:

> I alone have built upward, have dared to challenge the wind, asserting Man's courage and determination to master nature... Only... I have had the moral courage to attempt to outstare nature. That is my sole reason for building this tower. Here on the surface of the globe I meet nature on her own terms, in the arena of her choice. If I fail, Man has no right to assert his innate superiority over the unreason of the natural world.

However his giant tower is toppled by erosion, and the wind then abates. Ballard's theme of nature exacting retribution against man's technological hubris is repeated in *The Drought* (1965). This time the sea extracts revenge for its pollution by the human race with the hero describing the disaster arising because 'millions of tons of highly reactive industrial wastes – unwanted petroleum fractions, contaminated catalysts and solvents – were still being vented into the sea, where they mingled with the wastes of atomic power stations and sewage schemes'. The result is a 'resilient mono-molecular film' being formed over the oceans which stops evaporation and hence rainfall. The world suffers a global drought, food failures and collapse of civilization. The few survivors revert to a primitive state, scavenging for water; eventually rain does fall and the book ends.

Ecotopia

A big influence on the pastoral ideal was the rediscovery (and reprinting) in the early 1970s of many anarchist utopian classics such as *News from Nowhere* and *Fields, Factories and Workshops Tomorrow* (a new edition with an Introduction by the anarchist writer Colin Ward was published in 1974). With the publication of *The Dispossessed* by Le Guin in 1974, a whole new genre of SF was started about ecological, and generally anarchist, societies – to be labelled 'ecotopia', particularly after the publication in 1978 of a novel of that name by Ernest Callenbach.

The Dispossessed has been widely acclaimed by feminist and SF critics, and been much discussed in academic literature. The novel's original subtitle was *An Ambiguous Utopia* and its literary strength lies both in its discussion of traditional utopian themes of self and community, poverty and wealth, socialism and capitalism, and its attention to the dilemmas in the structures of the ideal state. Le Guin's ambiguity to this utopia is reflected in her short story *The Day Before the Revolution* (1974), which can be considered as an introduction to the anarchist society of *The Dispossessed*, being the tired, unromantic last memories of that society's founder.

Although not a child of the Sixties (she was born in 1929) her earlier work – such as *The Word for World is Forest*, written in the winter of 1968 but not published until 1972 – was strongly influenced by her active protest against the Vietnam War and by Tolkien's *The Hobbit*.[25] In *The Word for World is Forest* a planet is inhabited by small, furry, peaceful creatures, the Athsheans, who live in ecological and feminist communities. However (shades of Vietnam) their world is being colonized by humans, who calls them 'creechies' and treat them as slaves. Any who resist are hunted down, and their villages are attacked by napalm. Eventually the Athsheans launch a surprise attack and wipe out most of the human colonists who then retreat from the planet.

Le Guin's father was Dr Alfred Kroeber, a celebrated anthropologist on Native Americans, and this influence shows through clearly in her later work. *Always Coming Home* (1985) describes a matriarchal and tribal society in a post-holocaust California. Le Guin's influence on SF writers is immense and is acknowledged by Kim Stanley Robinson, whose work explores both the traditional SF themes of post-holocaust futures – as in *The Wild Shore* (1984) – and the colonization of space

(The Mars trilogy (1992, 1996)). Kim Stanley Robinson's search for an ecological and sustainable society which he terms 'Future primitive', incorporating new technology with primitive social forms of organization, draws heavily on the work of Le Guin and also of Ernest Callenbach to whom (along with the eco-poet Gary Snyder) he dedicates his book, *Future Primitive: The New Ecotopias*. As Robinson explained in this book:

> Le Guin is a crucial figure to the concept of Future Primitive. Her novel *Always Coming Home* tells the story of a far future that has reinvented Californian Indian culture on its own soil; it is a tribute to her father Alfred Kroeber's anthropological work in both form and content, and the fullest description of a postmodern/pale-olithic society I know. Le Guin's anarchist utopia *The Dispossessed* is one of the most important utopias of our time, and her Hainish novels, especially *The Left Hand of Darkness*, all depict worlds that are both primitive and starfaring.

Creating a sustainable future

SF can not only reflect current anxieties in society but also, in its utopian form, is a way of inventing a new consensus vision of what the future might be, not just technologically but also socially. Kim Stanley Robinson warned that we are in a race to invent and practise a sustainable mode of life before catastrophe strikes us and that SF is part of this work. He remarked in his book *Future Primitive*:

> All manners of alternative futures are now being imagined, and many of them invoke the wilderness, and moments of our distant past, envisioning futures that from the viewpoint of the industrial model look 'primitive'. It's not that they advocate a simple return to nature, or a rejection of technology… Rather, they attempt to imagine sophisticated new technologies combined with habits saved or reinvented from our deep past, with the notion that prehistoric cultures were critical in making us what we are, and knew things about our relationships to the world that we should not forget. These science fictions reject the inevitability of the machine future, and ask again the old questions, What is the healthiest way to live? What is the most beautiful? Their answers cobble together aspects of the post-modern and the pale-olithic… These visions are utopian statements of desire, full of joy and hope and danger, re-opening our notion of the future to a whole range of wild possibilities.

However SF is not only a medium for utopian visions but also for satire on those visions. Ecological communities make an easy target, and Stanislaw Lem, one of SF's leading satirists, in his *Star Diaries* (1976) has a community reject nuclear power in favour of electric eels as their source of energy. He mocks environmentalists who believe in unlimited possibilities of 'clean' technology as well as those who believe in technological progress as a source of human happiness.

Thus SF literature has reflected the yearning and ambiguity associated in Western society with a return to a pristine paradise or primitive past. As Richard Jefferies acknowledged in *After London* (1885), a simple rejection of industrialized society for a rural past is not a desirable answer, and Kim Stanley Robinson's attempts a century later to map out the path of a future society – albeit a Californian one – is laden with ambiguity over the use of technology. In his aptly named book *Future Imperfect: The Mixed Blessings of Technology in America* (1994), Howard Segal, a social critic of technology, remarked: 'To contemplate a return to a primitive paradise without technology – an attraction to some simplistic contemporary critics of technology – is thus to indulge in a historical fantasy no better than that of the technological utopians'.

Therefore the best (ecological) SF does not propose escapism from technology and its problems but rather nostalgia for a less technological society, and above all it raises the question of what kind of technological society we want to live in. This is more a question of human nature than of technology.

Chapter 7

SF's ecological vision

Few SF writers concerned themselves with agricultural problems. The countryside was romanticized rather than analysed. One of the few writers to examine agricultural issues, in the days of its great depression in the 1920s and 1930s, was Arthur George Street. He was well-known since the early 1930s as an author, lecturer and broadcaster on the English countryside and was active in supporting the organic movement. A. G. Street, perhaps influenced by the warnings of Sir George Stapledon's major study of land resources *The Land: Now and Tomorrow* (1935), wrote *Already Walks Tomorrow* (1938), which is an early novel of future ecodisaster. The novel warns of England's virtual starvation during the 1940s due to bad farming practices, and it is accurate and realistic in its account of the British and worldwide agricultural depression after 1918. The plot follows the fortunes of its hero, James Brockway, a Wiltshire farmer forced from his farm by the depression to become a frustrated civil servant with the Ministry of Agriculture.[26] A Lord Ellwood (modelled perhaps on Viscount Lymington[27]) encourages Brockway in his ideas, and sends him on a worldwide tour to report on soil erosion and the loss of soil fertility. Lord Ellwood recruits the support of Adam Read, a rich industrialist, to finance the reclamation of derelict farms run by Brockway according to his (organic) ideas. A worldwide famine strikes in the early 1940s due to soil exhaustion, erosion and drought but the trio of Read, Ellwood and Brockway saves Britain from famine. The worldwide crisis ends in 1947 and the book concludes happily with Brockway able to repurchase his family farm, and marry (after the fortunate death of his suburban wife) his soul mate Lady Ann Landon (who is Lord Ellwood's daughter).

What is striking about this book is the amount of detail on agricultural practice and policies. As Chris Morgan remarked, 'the author is giving his own recommendations here, based on many years' farming experience. It is not always enough just to warn; one must provide answers and show them succeeding, too'. Street's solutions are similar to those advocated by Stapledon, who in turn influenced the policies of the Kinship in Husbandry group, of which Viscount Lymington was a leading member. The novel's solution in terms of a revitalized Britain based on agriculture was a return to mixed traditional farming with crop rotation, and with land nationalization to protect against land abuse with farms on five-year leases. It also included social and welfare measures to stimulate rural life and urban consumption such as minimum wages and family allowances, birth control, and currency reform – ideas put forward in the 1920s by Guild Socialism and in the 1930s by Social Credit who were fringe political groups with strong links to the organic movement.

Already Walks Tomorrow glorifies the pre-war way of rural life, albeit from a landowning and aristocratic perspective which still persists today in supporters of the organic tradition such as Prince Charles. Street's views (and most pre-war conservationists) are summed up in the novel by the hero saying that 'care of the land is a national responsibility. Not until the land belongs to the nation can the nation be expected to farm it generously and well'. With this care based on Biblical ideas of stewardship 'men come and go, but the land remains, and its needs must ever be served'. This Christian passiveness in the face of nature would be reflected in post-war SF, especially the pastoral genre beginning with *Earth Abides* (1949) which is dedicated to the verse from Ecclesiastes (1: 4), 'Men go and come, but earth abides'.

The problems of worldwide soil erosion were however brought to a much wider audience by the publication of the (eventually) widely selling book *The Rape of the Earth* (1939) by Graham Jacks and R. O. Whyte. There was a continuous stream of books on agriculture written from the organic perspective during the 1940s both in Britain and the USA, but ultimately these were in vain as post-war agriculture chose the path of increased industrialization, reinforced by the introduction of new pesticides and insecticides such as DDT.

Street was not the only agricultural writer to use the SF genre to communicate the organic message. This was also done in Edward Hyams' satirical novel *The Astrologer* (1950) that praises the maintenance of soil fertility. The novel's hero exclaims: 'Soil... What we live on. It's running out. The nine-inch outer skin of the planet, the sole source of terrestrial life and we've nearly used it up. I made a lot of calculations about a month ago. Universal famine, fighting for scraps of food, mutual slaughter, plague, revolution and death. That is the prospect'. This novel was published two years before Hyams' non-fiction work *Soil and Civilization* (1952) that attacked the mechanization of agriculture.

The ecology of *Earth Abides*

After the dropping of the atom bomb in 1945 the tenor of most SF novels was pessimistic, and they delighted in showing a devastated Earth without much hope of the survival of mankind. From the end of the 1940s to the late 1960s, SF mainly concerned itself with dystopian visions of despotism and disaster, and myths of rebirth in which the few remaining survivors rebuild the good society. In the same year, 1949, that were published the visions of society destroyed, in *Ape and Essence* by Aldous Huxley and *Nineteen Eight-Four* by George Orwell, came a novel to give some grounds for optimism in a post-Hiroshima era. This was *Earth Abides* by George Stewart, an English professor at Berkeley, and its message was that mankind did have a future if it was willing to live in harmony with nature.

The theme of this very popular novel is the insignificance to Earth of the human race and civilization, with strong Biblical overtones.[28] It also has explicit discussion of ecological concepts, as the narrator, Isherwood Williams or 'Ish', is a biologist and (at the start of the book) a graduate student doing his book on 'The Ecology of the Black Creek Area'. As civilization and technological systems decay he muses on (and gives examples of) ecological succession, the survival of the fittest, the competition between species, population instability, and the permanence of nature. Foreshadowing deep ecology sentiment he reflects on how little difference human extinction would make to the larger rhythms of life on earth. Ish believes civilization deserves to vanish: 'Perhaps there were too many people, too many old ways of thinking, too many books... Why should not the philosopher welcome the wiping-out of it all and a new start and men playing the game with fresh rules! There would be, perhaps, more gain than loss'. Thinking about the relationship between man and nature he declared: 'In the times of civilization men had really thought themselves as the masters of creation. Everything had been good or bad in relation to man... But now Nature had become so overwhelming that any attempt at its control was merely outside anyone's circle of thought. You lived as part of it, not as its dominating power'. Finally in old age, as his beloved university library crumbles away with its stock of human knowledge, to be lost forever, he ponders man's yearning for escape from civilization to some simple life. He thinks: 'Thoreau and Gauguin – we remember them... What a strange thing then is this great civilization, that no sooner have men attained it than they seek to flee from it!... Why do legends look back towards some golden day of simplicity?' As his small community is forced back to the 'simple life' of hunting and gathering, and reverts to the pre-European life of American Indians, he becomes reconciled to the loss of civilization and knowledge. As he dies content, surrounded by his grandsons, gazing on the rusting towers of the Golden Gate Bridge, the book ends: '"They will commit me to the earth," he thought. "Yet I also commit them to the earth. There is nothing else by which men live. *Men go and come, but earth abides*"'.

Stewart's explicit use of ecological concepts in a SF novel was pioneering, and undoubtedly reinforced a favourable climate amongst its many readers for ecological literature, coming at the same time as the publication of *Our Plundered Planet* by Fairfield Osborn and *The Road to Survival* by William Vogt. These two popular non-fictional works portrayed a world on the edge of catastrophe from over-population and resource shortages. Ecologists were now not only 'prophets of doom', but also advocates for a new way of life that looked not to nostalgia for a rural past but the role of ecology in a viable future.

The conservation message of Frank Herbert

Another SF writer who used ecological concepts was Frank Herbert who was to achieve cult status with his bestselling *Dune* (1965), with its mythic battles for survival of a planet's inhabitant, the Fremen, against alien invaders and a hostile environment. One of the central characters and heroes is the Planetary Ecologist. and *Dune*'s impact lay in its treatment of ecology, a theme which it brought into the forefront of modern SF readers' and writers' awareness. As Malcolm Edwards remarked in *The Encyclopedia of Science Fiction*, the Fremen 'clinging to the most precarious of ecological niches through fanatical scrupulousness in water conservation, is possibly the most convincing alien environment created by any SF writer'.

Not so well-known is Herbert's *Green Brain* (1966) whose plot is about an attempt to exterminate insect life in the Amazon through the use of pesticides. The result is the creation of super-insects and a controlling Brain – a common SF concept. This novel is heavily influenced by the contemporary debate over pesticide use in the West – the group campaigning against insecticide use are called 'Carsonites' – and attempts by China to wage war on nature and, during the 'Great Leap Forward', to fanatically eliminate flies and mosquitoes. The novel warns against the dangers in opposing nature, and is a plea for the interdependence of life and what we now call 'biodiversity'. At the end of the novel the insect Brain says to its captured humans, 'Break the chain and all die. The more different forms of life there are, the more life the greenhouse can support. The successful greenhouse must enclose many forms of life – the more forms of life, the healthier for all'.

The Urban dystopia

Before 1945 few SF writers concerned themselves with the impact of population growth on human society. One of the few who did was W. D. Hay in his novel *Three Hundred Years Hence* (1881) that imagines a world of 130 billion people living in cramped but luxurious apartments in giant underground cities. The surface of the earth is given over to hi-tech agriculture with most wild animals exterminated, and the few remaining kept in zoos. This image of humanity living in crowded hi-tech cities in an over-populated world devoid of wildlife and nature was to become the staple of post-war SF, this time as dystopia in such bestselling SF novels as *The Space Merchants* (1953), *The Caves of Steel* (1954), and *Make Room! Make Room!* (1966).

The Space Merchants (1953) by two well-known SF writers, Frederick Pohl and C. M. Kornbluth, sold over 10 million copies in 40 languages. In this satire the advertising industry runs the world through puppet governments, and rival firms battle each other (sometimes violently) for world domination of consumer markets through brands. Population growth is considered essential for creating new markets, so the world is over-populated, resources are scarce and cities very crowded. The only opposition is from the radical group 'The World Conservation Association' (W.C.A.) popularly known as 'the Consies'. They operate as a classic guerrilla group in secret cells and engage in sabotage and graffiti; hence they are labelled by the advertisers as 'Consie saboteurs'.

The authors make explicit reference to the works of Osborn and Vogt by referring to 'the semi-mythical Conservationist heroes, Vogt and Osborn, whose books were required reading in all cells'. The hero Mitchell Courtenay, once a high ranking advertiser, has fallen victim to a rival firm's machinations and been reduced to near slavery producing synthetic food on a plantation in Central America. There he is recruited by the Consies and given a leaflet that says:

> The Facts: The W.C.A. is a secret organization persecuted by all the governments of the world. It believes that reckless exploitation of natural resources has created needless poverty and needless human misery. It believes that continued exploitation will mean the end of human life on Earth. It believes that this trend may be reversed if the people of the Earth can be educated to the point where they will demand planning of population, reforestation, soil-building, deurbanization, and an end to the wasteful production of gadgets and proprietary foods for which there is no natural demand. This educational programme is being carried on by propaganda – like this – demonstrations of force, and sabotage of factories which produce trivia.

Courtenay goes along with the Consies in order to escape back to New York, where after various adventures, including on the Moon, he manages to outwit his enemies, reclaim his old job (and win back Kathy, his ex-wife). He arranges for the Consies to achieve their goal of emigrating to unspoiled Venus, to create a pastoral paradise.

Perhaps the success of this book is that it mixes the two popular post-war themes, firstly of control by corporations and advertisers – which was much discussed by social critics such as William Whyte in *The Organization Man* (1956) and by Vance Packard in *The Hidden Persuaders* (1957) – with neo-Malthusian warnings popularized by Fairfield Osborn, William Vogt and Harrison Brown. An enormous success, it pioneered the dystopian vision of the over-populated city running out of resources, with people living at best in tiny apartments (or often on the street) on synthetic food. It was a vision of the teaming poverty of the Third World (Indian) city, of rickshaws, street markets and public water taps, crossed with the affluent lifestyle of corporate New York, of country clubs, luxurious living and private security guards.

Asimov's optimism

An alternative vision of an over-populated city is *The Caves of Steel* (1954) by Isaac Asimov, one of the few technological optimists writing in post-war SF. This time we have a vision of a world of robots and space colonies. In New York city millions live under a giant Dome, isolated from, and fearful of, the countryside outside the walls, which they refer to as 'the wilderness' and never enter. It is a world of food rationing, of synthetic food, and resource shortages particularly of uranium to run the nuclear reactors that power the city. Again there is internal opposition, this time from a loosely organized secretive group labelled 'the Medievalists' who are nostalgic for the old way of life outside the cities, and who oppose contact by 'the Modernists' with 'the Spacers', the hi-tech space colonists. A secret sympathizer of the Medievalists describes them: 'They lived closer to nature. It's healthier, better. The troubles of modern life come from being divorced from nature'. However the hero, Lije Baley, is a skeptical person and unimpressed as he believes that moaning about technology going wrong was 'a built-in facet of human nature'.

Baley discusses with Norris, a colleague, the Modernists' dreams of overcoming the Earth's energy shortages by harnessing the unlimited solar energy in space, and transmitting it to earth via an energy beam. Anticipating what technological optimists Julian Simon and Herman Kahn would preach nearly three decades later in their book *The Resourceful Earth* (1984), Norris says: 'There's no way you can stop mankind, Lije. You've got to be optimistic about it and have faith in the old human brain. Our greatest resource is ingenuity and we'll never run out of that, Lije… When it's necessary, it'll be done. Why worry?'

Reflecting on this, Baley has a vision of Earth with unlimited energy and how this would solve all resource problems. He concludes: 'There was a time when the current population of eight billion would have been viewed as impossible. There was a time when a population of a single billion would have been unthinkable. There have always been prophets of Malthusian doom in every generation since

Medieval times and they had always proven wrong'. The novel ends optimistically with the possibility of future generations leaving Earth to become space colonists. In a sense the conclusion is the same as *The Space Merchants*, that of the possibility for new beginnings for the human race (and of founding utopian societies) through the age-old device (then very popular in SF) of colonizing new worlds. However a decade later the difficulties of space travel had become apparent, and escape from Earth was no longer considered a solution.

The bleakest vision

The bleakest Malthusian vision of an over-populated New York is found in *Make Room! Make Room!* (1966) by Harry Harrison which uses much of the imagery of Ballard's earlier short story *Billenium* (1961). Harrison's book is explicitly intended to convey an ecological message on the dangers of over-population and the need for birth control. As Harrison said in *Future Imperfect*: 'It was written at a time when there were no popular books about population. I did a bit of work – three, four, five years of research. And I wrote this novel in a world not generated by me. I used everybody else's facts. I used the worst case scenario. This is what science fiction does; it shakes the admonitory finger. Shape up or this is what you can expect'.

The novel is set in New York in 1999, vastly over-crowded with a population of 34 million people as birth control is banned. Following the themes of the dystopian city of the future, natural resources such as water, fuel and food are scarce or non-existent, and people live on synthetic food made from seaweed. New York has been reduced to a Third World city, with people living on the streets, there are no cars (as fossil fuels are exhausted), only pedicabs (rickshaws). The masses scavenge and subsist on welfare handouts, while a small elite lives in luxury. Life is full of corruption, squalor, and brutality. The hero is Andy, a policeman who shares a room with Sol, an old man who can remember the days of plenty but has sunk into apathy. The message of the novel is given by Sol who, near the end of the book, makes an impassioned plea for birth control, criticizing the Catholic Church and other reactionaries who oppose it. Sol exclaims: 'So mankind gobbled in a century all the world's resources that had taken millions of years to store up, and no one on the top gave a damn or listened to all the voices that were trying to warn them, they just let us overproduce and overconsume, until now the oil is gone, the topsoil depleted and washed away, the trees chopped down, the animals extinct, the earth poisoned, and all we have to show for this is seven billion people fighting over the scraps that are left, living a miserable existence – and still breeding without control. So I say the time has come to stand up and be counted'.

Sol then goes out to join a demonstration in favour of birth control, but gets injured and dies due to lack of medical attention. The novel has a very pessimistic ending, with Andy unjustly demoted from his job and Shirl, his girl friend, leaving him because she can't stand the lack of privacy in their room which they have to share with a large family. The novel concludes with Andy on New Year's Eve 1999 in Times Square, looking at a giant TV screen that says, '344 Million Citizens in these United States. Happy New Century! Happy New Year!'

This novel has the bleakest ending I have come across in SF: there is no hope for the individual in an overcrowded and brutish world. Harrison deliberately made it so. Despite its pessimism the novel sold well and was filmed as 'Soylent Green'. It came in an era of international concern and media attention over the dangers of over-population, and national debates on the legalization of birth control and abortion. However this is no ordinary SF book. It is similar to the popular paperback *The Population Bomb* by Paul Ehrlich that was to follow two years later. Harrison's book starts with a Prologue containing a statement by President Eisenhower on US government population policy and ends with a two-page reading list of books and journals. About half are by social and political writers such as Erich Fromm, Jane Jacobs, Wright Mills and Vance Packard and the other half by ecological writers such as Harrison Brown, Charles Elton, Fairfield Osborn and William Vogt.

Harrison's novel, like Ehrlich's book, was intended to inspire people to action through creating a vision of an over-populated dystopian world, and both books were intended to engage the reader's emotions with doom-laden tales. However SF is commercial publishing aiming for large sales, and most writers will use currently popular themes to construct an exciting narrative of events in the future. Thus in the mid-1960s with public attention on the over-population problem it became a natural topic for SF writers, and novels with this theme came from well-known writers. They included John Brunner's *Stand on Zanzibar* (1968), and James Blish and Norman Knight's *A Torrent of Faces* (1968).

In conclusion, SF reflects the social concerns and the scientific debates of the day and strongly interacts with the scientific views of the future, as put forward by mainstream scientists and writers. It is a powerful form of propaganda for competing visions of the future and, I believe, a formative influence on young minds and the literature of modern environmentalism.

Chapter 8

Atomic energy in the 1960s

The official histories written from both the pro- and the anti-nuclear viewpoint both give common explanations about how and why, and indeed when, the anti-nuclear movement emerged. But these explanations are being challenged as myths by a new generation of researchers, particularly Ian Welsh in his book *Mobilising Modernity*. The traditional explanations are:

- There was only opposition to nuclear power since the early 1970s.
- Prior to this there was a 'golden age of public acceptance'.
- Friends of the Earth (FoE) started and led the anti-nuclear movement.
- The activities of the anti-nuclear and environmental movements led to nuclear power's demise.

Similarly O'Riordan, in the journal *Catalyst* in 1986, divided the life cycle of civil nuclear power in the UK into four distinct ages, the first being 'the age of innocent expectation' from 1946 to 1966, years which he claims were characterized by innocence and euphoria, when powerful men backed by powerful organizations ruled with a combination of arrogance and complacency. This was followed by 'the age of doubt' from 1967 to 1974, 'the age of anguish' from 1975 to 1980, and 'the age of public justification' from 1981 onwards.

This chapter and the next challenge the conventional view that practically all anti-nuclear opposition before the late 1970s was expressed within the public inquiry system in Britain, and that there were no demonstrations, pickets, or any other signs of protest.

Nuclear euphoria in Britain

Until the early 1970s criticism of nuclear power was restricted to specialist journals, elite newspapers (generally *The Times*), parliament and industry forums. In Britain the 1955 White Paper *A Programme of Nuclear Energy*, on the first nuclear power programme, was greeted with acclaim by the press, which followed it up with glowing reports on the triumphal opening of Calder Hall by the Queen in October 1956. After the Suez Crisis in 1956, the government took the opportunity to triple the expansion of the nuclear industry. According to O'Riordan nearly all the establishment and the media saw nuclear power as part of the 'white heat of an oil-free technology that would bring Britain into a new age of prosperity'.

However expression of nuclear optimism from within scientific and political circles did not go completely uncriticized according to Ian Welsh. A small group of parliamentary backbenchers raised a series of issues, including reactor safety, the adequacy of plans for nuclear waste disposal, the dangers of low-level radiation and the absence of any integrated energy policy. In response the government was 'perplexed' at the expression of 'anxieties which it is not usual to hear'.

Both *The Times* and *The Economist* were critical of the tripling of the nuclear programme, calling it too ambitious. *The Times* commented in an editorial on 6 February 1957 that 'Atomic energy has the power to evoke fantasies. It is not only looked on as an "answer to Suez"... but as a fairy godmother source of cheap electricity'. It further cautioned against prophets who visualized that in twenty years time there would be 'electricity on tap in every house, like water, laid on for a low fixed charge irrespective of the quantities consumed'. Unswayed by the myth of cheap atomic energy, *The Times* warned that 'The hard fact is that electricity from the first atomic power stations will be dearer, not cheaper, to generate than power made in new coal-based stations'.

In conclusion *The Times* argued in favour of a more balanced approach to energy, including the underground gasification of coal, rather than a massive financial commitment to nuclear. This latter criticism was echoed by *The Economist* (19 January 1957) which protested against 'the immense capital expenditure' before 'the first atomic power stations had time to prove themselves'. However such concerns could scarcely be heard within the overall climate of adulation.

The Stretch critique

One vocal critic from within the AEA was Kenneth Stretch, works manager at Calder Hall from 1954-7, who argued against the dominance of theoretical work as a guide to policy development. In the journal *Nuclear Power* in December 1958, he put forward a rare and powerful critique of British nuclear economics. Stretch, commenting on a lecture the previous year in Sweden by Lord Hinton (the newly appointed chairman of the CEGB), questioned both Hinton's cost forecasts for nuclear electricity and the wisdom of so large a commitment to one reactor system. He believed that Hinton's figures were 'seriously biased in favour of nuclear power' and he did not expect nuclear power to be competitive with coal until 1980-90. It was therefore 'extremely rash for this country to rush into building a large number of obsolescent nuclear stations' and over-optimistic forecasts of nuclear costs were 'likely to rebound very awkwardly'.

Although Stretch's paper was published in a prominent journal and was drawn to the attention of ministers in parliament, it was ignored by both Lord Hinton, who never troubled to read it, and the AEA. Stretch, according to Welsh, was harshly treated for his criticism; he was ostracized by his colleagues and became known in the AEA as 'The Renegade'. He eventually resigned to teach engineering in Birmingham, and he further expanded his ideas on the nuclear power industry and failings of modern science in a book published in 1961. Lord Hinton only acknowledged the truth of Stretch's critique in 1975, admitting that it was an 'over ambitious nuclear programme' in a retrospective review of nuclear policy in a House of Lords debate, and in an article published in *New Scientist*, in October 1976.

Fuel policy in the 1960s

Most criticism in the 1960s was of the decisions made on the speed, scale and reactor choice of the nuclear programme, not of the nuclear project itself. In the early 1960s the AEA forecasts were quickly proved in several respects to be overoptimistic: the Magnox reactors took longer and cost more to build than expected, while the cost of power from coal fell. The formation of OPEC in 1960 was generally dismissed, except by Fritz Schumacher at the NCB who continually prophesied on the dangers of oil dependency.

The secrecy of the deliberations on reactor choice by the Powell Committee had by 1964 brought criticisms of excessive nuclear secrecy from the press. The opinion of *Nature* (27 June 1964) was not untypical: 'The secrecy maintained over the proceedings of the Powell Committee is indefensible'. The *New Scientist* (19 March 1964) complained that 'poorer decisions were to expected from a secret technocracy than from a more open policy system'.

The nuclear programme attracted criticism from academics of the period, who were very critical of the size of the first nuclear programme, and of the decision to concentrate on one reactor type alone. Typical of such criticism was the work of Mary Goldring who, in the *Journal of the British Nuclear Energy Society* in January 1964, wrote that the worst mistakes of the nuclear power programme were its rigidity and inflexibility. She warned: 'Too much atomic planning is done today by people pushing atomic power', and that cost estimates were 'games played by accountants'.

However the decision to choose the AGR, rather than the LWR, for Dungeness brought rave reviews for the nuclear industry, with the British press giving the AGR's success an enthusiastic and mostly uncritical reception, calling it a 'triumph' and a 'breakthrough'. Few outside the technical community recognized, however, that it was on the basis of the CEGB's assumptions and not on basic technological and economic merit that the AGR had won. One vocal critic of the AGR decision was Duncan Burn, whose 1967 book *The Political Economy of Nuclear Energy* criticizing the AEA's monopoly over R&D caused a sensation and prompted many influential personalities within the industry to commit themselves either to Burn's attitudes or to the defence, as they saw it, of the Atomic Energy Authority.

The years 1965-68 had two White Papers on Fuel Policy, in 1965 and 1967, accompanied by much internal fighting between the coal and nuclear industries, and extensive newspaper coverage on the (bleak) future for coal and (rosy) prospects for nuclear. The most organized opposition to nuclear was from the coal industry led by Lord Robens,[29] the chairman of the NCB. He was described by R. F. Pocock, a contemporary nuclear historian, in his book *Nuclear Power* (1977) as 'one of the most severe critics of the nuclear power programme, which he saw as the principal threat to the mining industry' and 'an outspoken man and formidable opponent'. From early 1966 Robens was in continual conflict with Richard Marsh, the Minster of Power who had a similar reputation. Such was the threat to coal from nuclear that the NCB took the highly unusual step of effectively dissociating itself from the government's fuel policy in its Annual Report for 1966/67. It argued:

> Estimates that the cost of providing electricity from nuclear stations will be less than the cost of doing so from thermal stations need close and dispassionate examination. Nuclear stations cost huge sums in capital and account needs to be taken of the use to which excess capital expenditure could be put if it were not sunk in these stations. To the extent that the basic technology remains unproven, the justification for spending these large sums of capital must be open to question.

The 1967 White Paper was meant to resolve these economic questions but as Robens commented in his biography *Ten Year Stint* (1972): 'No doubt the Labour Government though that when the White Paper was published in 1967 the public debate about the relative merits of nuclear power and coal would come to an end. On the contrary, I continued to clamour for an independent inquiry into these matters'. In desperation the NCB commissioned a report from the Economist Intelligence Unit, which was published in September 1968, but the Ministry of

Power rejected its 'middle of the road policy'. A further report critical of government policy came in 1968 from the Brookings Institution, which concluded that 'Coal output should probably be cut back more slowly than is planned and the investment programme for electricity (especially nuclear generation) should be substantially reduced'. But overall nuclear had a very good press in the 1950s and 1960s, and the few critics such as Stretch, Burn and the NCB were easily dismissed.

The 1950s public inquiries

Between 1956 and 1961 there were seven public inquiries into proposals to build Magnox stations. The first (commercial) Magnox, at Berkeley in 1955, was approved without an inquiry but from then on every Magnox, except Sizewell A in 1960, was subject to an inquiry lasting from two to five days. This compares to all five AGRs built without an inquiry in the 1960s. In fact there was a ten-year interval (1961-71) without an inquiry. The 1970s saw a return to nuclear inquires. In 1971 there were two: the first over Connah's Quay lasted 28 days during March and April, and the second over Portskewett for three days in June. Then in 1974 there was an eight-day inquiry over Torness followed by the marathon hundred-day inquiry over the Windscale THORP plan in 1977.

Brian Rome, a law lecturer and Conservation Society activist at Bristol Polytechnic in the 1970s and 80s, compiled extensive indexes of all nuclear power inquiries. Writing in his 1979 *Index to Past Public Inquiries*, he believed that past inquiries contain much information relevant for forthcoming ones. He provides references to show that the issues of need, non-nuclear alternatives, reactor safety, emissions, economics, and accidents that dominated the 1970s anti-nuclear campaign had already been raised in the 1950s. The need for nuclear power was questioned at the Bradwell inquiry in 1956, at Hinkley in 1957, at Trawsfynydd in 1958 and Wylfa in 1961. Reactor safety was an issue at Bradwell, Trawsfynydd, Dungeness and Oldbury, while economics was raised at Bradwell, Oldbury and Wylfa. As the public inquiries progressed the range of objections widened from initial concerns on amenity, nature conservation and economic interest to a far more sustained attack on the safety and economics of nuclear power as occurred at the Oldbury inquiry, held in April and May 1960. However national press coverage of the inquiries was mostly confined to short reports or letters, mostly on wildlife or amenity issues, with local press giving more space to opposition groups.

Amenity issues were central to objectors' arguments and press coverage, and reflected the fact that amenity was a familiar reason for dissent. In the prevailing language of the 1950s amenity had a similar connotation to NIMBY in more recent times. The building of large industrial plants in rural areas was considered a threat to England's 'green and pleasant land' and found expression in the House of Lords, the press and at the inquiry. The amenity issue was especially important in two location decisions, the siting of twin Magnox reactors at Trawsfynydd within the Snowdonia National Park and at Dungeness. These developments achieved a relatively high public profile due to the outstanding beauty of these areas and the efforts of naturalists in opposing the developments.

Bradwell

The first public inquiry into a nuclear power station occurred at Bradwell-on-Sea, Essex in April 1956 and is covered in detail by Luckin (1990) and by Welsh (2000). Initially Aubrey Jones, the Minister of Fuel, hoped it might be possible to avoid an inquiry but opponents of the scheme exploited the contradictions in the siting criteria. If nuclear was as safe as conventional industries why must it be placed in remote rural areas? This question of why nuclear stations were located in remote areas if they were so safe was to bedevil all future debate over nuclear safety. The local MP, Brian Harrison, had taken informal soundings at the beginning of 1956 and found that the safety issue had been 'worrying local residents more than anything'. Although a large number of villagers, at a parish gathering in March, declared themselves in favour of the scheme, the opponents, who included national figures such as Tom Driberg, ex-MP for the area, and John Betjeman, a conservationist for Victorian architecture and later to become Poet Laureate, were better organized and better able to influence public opinion through use of local media. The opposition formed the Blackwater and Dengie Peninsula Protection Association (BDPPA) with the intention of creating the strongest opposition to the project by running a 'campaign to excite national interest'. By mid-March the Minister of Fuel had received more than 150 formal objections, and therefore had little option but to sanction a public inquiry. The BDPPA reflected two sets of organized local interests centred around the oyster industry and local sailing enthusiasts and acted as a focus around which a much more diverse and less clearly focused range of objections coalesced, many from outside the local area.

The inquiry lasted five days in April and May and was, as the inspector noted, well attended by the public with objections received from as far afield as Somerset and the Midlands. The inspector however judged the evidence submitted purely in terms of expert testimony on amenity, the alleged danger to the oyster fisheries, and the toredo worm. The everyday fears and suspicions of the objectors were dismissed as 'unreal'. This left only the testimony of the two expert witnesses for the objectors on the possible damage to the oyster beds and the prospect of an increase in 'toredo worms' and 'gribble' which were detrimental to yacht hulls. According to Welsh the inspector dispatched their evidence by recourse to a childhood reminiscence of a summer spent watching the oyster fishermen of Whitstable. From his extensive study, the inspector concluded that 'oyster culture was a very chancy business' and that 'one cannot always account for a bad harvest'. Finally the inspector dismissed the amenity issue which was 'an emotional one upon which the whole opposition case hinged' by saying that amenity must bow to the face of progress and the inevitability of change which was already occurring.

The coverage by the national media was patchy, and there were no in-depth feature articles as there had been for the Magnox reactors at Calder Hall. Objectors' fears about nuclear safety went unreported and the press tended to follow the prejudices of the inspector in its reporting, with often inane headlines about 'worms'. The Minster of Fuel, Aubrey Jones, under pressure to keep to a demanding nuclear timetable, accepted without qualification both his inspector's report and his dismissal of the objectors' case. Objectors were left to plead their case through the letters columns of the national press. The objectors wrote to every MP arguing that their case had not been given adequate consideration and urging that the inquiry be re-opened. Their plea fell on barren ground.

Hunterston

The inquiry into the proposal to build a Magnox station at Hunterston, on the Clyde in Scotland, lasted 11 days, and was held in January and February 1957. It was re-opened briefly on June 14th to enable an inspector's report to be submitted as the original inspector had died in the intervening period. The proceedings were published as a White Paper by the Scottish Office in 1957. Objections were received from forty individuals and organizations. These included an individual from Kent and a petition bearing 208 signatures. Unlike Bradwell no commercial interests were represented. In this case the major objector was the Hunterston Estate upon whose land the station was to be sited. Objections covered both nuclear and non-nuclear concerns. The former were about the 'effect of radioactive dust and fumes' and their implications for human life and the flora and fauna of the area. The latter covered the loss of agricultural land; the presence of alternative sites; damage to roads; disturbance; long-term economic impact due to imbalances created within the local labour market; the impact upon fisheries; and finally, amenity considerations.

The nuclear establishment tried to address public opposition over what they perceived as fear of an atomic explosion. As a result AEA spokesmen incessantly declared this impossible whilst insisting that the stations be sited away from population centres. As the *New Scientist* (28 February 1957) noted: 'This is done because of public nervousness: it seems a curious way of discouraging jumpiness'. Ian Welsh however believes that the authorities misread the pulse of public opinion by perceiving public concern to be based on the fear of atomic explosion. He commented:

> Whilst the fear of atomic explosion certainly existed, and was even fuelled by official reassurances, the effects of radiation were of equal significance. The continuation of atmospheric testing and presence of expert dissent around the effects of `fallout' which achieved a relatively high public profile were central to this. Throughout this period Hansard is peppered with questions relating to the genetic effects of radiation from atmospheric testing.

Furthermore he noted that there were parliamentary calls for inquiries into the foreseeable 'problem of disposal of atomic waste from power plants', but these were met with official silence or bland responses. During the inquiry it became known that the final contract for the station had already been awarded before the inquiry had even opened. This caused a furore that found its way to the House of Commons where it was described as 'the worst mistake' and gave the 'impression that the Authority was ready to ride roughshod over the rights of the individual'. Objectors felt that the outcome of the inquiry was a foregone conclusion. This feeling was intensified when the SSEB physically trespassed upon the estate to conduct preliminary site evaluation. This stirred up a 'great deal of ill feeling' and the inspector commented: 'Excess zeal does not atone for trespass'. Thus the Hunterston inquiry raised the argument, developed with much effect by Mike Flood and Robin Grove-White in their pamphlet *Nuclear Prospects* (1976), that the development of nuclear power represented a threat to individual liberty and human rights.

Trawsfynydd

The second round of sites selected for Magnox stations were located in areas where MPs and councillors welcomed the developments and the prospect of local opposition was minimized. However the selection of Trawsfynydd in the recently created Snowdonia National Park generated controversy. The locals were strongly in favour on employment grounds, whilst national conservation groups, which included the Ramblers Association, the Youth Hostels Association and most importantly the National Parks Commission, were opposed on 'amenity' grounds.

This involved not only dispute over the aesthetics – the modernist architecture of the station – but also its location in an area of outstanding natural beauty. An inquiry was held from 12-14 February 1958, with daily reports in *The Times*. The opening of the inquiry attracted a demonstration march of supporters, complete with banners, for which the CEGB was duly grateful. Delay in announcing the result caused local restlessness with *The Times* of 6 June 1958 reporting that local people showed eagerness to attract the project by calling for 'bread before beauty'.

The Times in its editorial said that 'the NFU branch complained that the National Parks Commission was retarding progressive development in the country generally, which was having a depressing effect on agriculture and economic development'. Furthermore it argued that 'the government was withholding permission for a decent interval to placate the amenity interests, whose arguments and interventions must be recognized as stemming from the most worthy motives'.

Dungeness

The argument that economic development should take precedence over amenity and wildlife issues had few critics in the late 1950s. One person who was prepared to take a stand was Max Nicholson, the director-general of the Nature Conservancy, an organization set up by the government in 1949 to protect wildlife sites. The proposal to build a Magnox station at the site of the Dungeness nature reserve brought an uncompromising and explicitly 'environmental' response from him.[30] At a press conference on 8 December 1958 reported in *The Times* the following day, Nicholson said he would oppose absolutely Dungeness power station at the inquiry. He said we must 'avoid the sacrifice for all time of by far the most important shingle structure in the British Isles and probably in Europe'. The Conservancy in its written evidence to the inquiry wrote that the land in question was of unique and irreplaceable permanent importance to science, while the nuclear power station was of relatively short-term utility. Furthermore in an attack on nuclear siting policy it declared that 'the long-term national economic interests which the programme of nuclear power development is designed to serve, are inconsistent with the selection of a site so far from the point of consumption'. It asked the inspector not to recommend the scheme.

At the inquiry, which lasted from 16 to 18 December 1958, the counsel for CEGB made a strong attack on the Nature Conservancy. He said that their evidence was 'intellectual opportunism', was full of exaggerated language and false suggestions, and was a lamentable performance by a public authority. Furthermore he said (as reported in *The Times* of 16 December 1958) that their evidence 'was obscure, and at worst it was equivocal, evasive and disingenuous. The arguments were irresponsible... How it justifies this ill-advised and ill-informed incursion into matters which are quite outside its functions is difficult to understand but it is disturbing to see'.

The CEGB put forward the contemporary view that the amount of damage done to minority interests would be relatively small by comparison with most sites that could be found. Two days later, in his evidence to the inquiry, Nicholson replied to his critics and said (*The Times* of 19 December 1958) that everybody connected with the inquiry had a 'split personality' for what was being discussed 'was a collision between the two personalities – we wanted economic development and a higher standard of living but we also wanted to leave something of our inheritance to our children'.

However his evidence was seen in scientific terms rather than ecologically ethical terms, and Dungeness was presented as a conflict between scientific research and economic development. The station was approved in June 1959 and as a result the Nature Conservancy abandoned its proposed Dungeness National Nature Reserve. Naturalist bodies expressed their disappointment, pointing out that this site had only been preserved through their efforts to maintain its wild character in perpetuity as a national asset. The chairman of the Kent Naturalists Trust wrote to *The Times* (15 July 1959) expressing his shock at the decision, saying that the site was unique in Europe and of great scientific value. He said: 'The irony of the situation is that in the name of the advance of science, the Authority is destroying an area which is far more valuable as a site for scientific study that as a site for a power station. The power station could go elsewhere but we cannot move the unique structure of Dungeness nor its wild life'.[31]

In other letters naturalists suggested a review of nuclear siting policy, with the construction of nuclear stations nearer to built-up areas, which they hoped would protect remaining undeveloped coastal areas. This echoed arguments put forward at the inquiry that if nuclear stations are safe, why build them in isolated spots? Ironically when the government did decide to change its siting policy in February 1968, and allow AGRs to be built much nearer to built-up areas than previously, one result was the support of the Nature Conservancy but strong opposition from local residents. This is what occurred at Stourport-on-Severn in early 1970.

Chapter 9

Schumacher's ethical campaign

This chapter looks at the anti-nuclear campaign waged by Alf Robens and Fritz Schumacher at the National Coal Board in the 1960s. The appointment of Alf Robens as chairman of the National Coal Board (NCB) in 1960 heralded the start of the first national campaign against nuclear power in Britain. As Robens wrote in his autobiography *Ten Year Stint* (1972), 'The battle of nuclear power versus coal began as far as I was concerned in October 1960, the month that I took up my post as Chairman Designate of the Coal Board'. His ten-year campaign during his tenure at the NCB was ably assisted by Fritz Schumacher, the chief economist of the NCB. In her biography *Alias Papa* Barbara Wood, Schumacher's eldest daughter, wrote: 'Robens was not the sort of man to preside over a declining industry. He understood at once it was important to keep the coal industry going and used Fritz to feed him with all the arguments he needed in the battle to keep the industry not only alive but also thriving. Robens did the fighting, Fritz supplied the ammunition'.

As Barbara Wood pointed out: 'The arguments with which Fritz furnished Robens, and which he himself put forward endlessly in lectures and articles, were basically the same as those he had put forward in the 1950s, only stronger. He pointed out the finite nature of the non-renewable energy resources and the foolishness of abandoning one major source just because another happened to be cheaper in the short term'. Robens was not anti-nuclear per se, and his campaign can be seen as a vested interest one. He, and the Coal Board as a whole, were not against the building of nuclear power stations, but argued that the proposed nuclear programme was too big, and that before proceeding to a large and (as it turned out) disastrous programme operating experience should first be gained. He wished to protect the coal industry from a too rapid shut down, and argued vociferously that nuclear power was being unfairly subsidized. Schumacher, in contrast, was against any nuclear power. Wood wrote of the NCB members:

> Their faith in nuclear energy horrified him more than anything. He had been collecting statistics about nuclear energy since his first meeting on the subject three days after joining the coal industry... There was no question in his mind that the impact it would have on the supply of energy in the 1960s, 1970s and 1980s would be negligible. Not only was it the wrong kind of energy, contributing only to base load electricity needs, but in percentage terms of total energy needs, its contribution was insignificant... To replace coal by nuclear energy just did not make economic sense.

The Hartlepool crisis

Robens' campaign to protect the coal industry and to expose the subsidies given to nuclear power had the backing of many Labour MPs, especially those from mining constituencies. However the Labour government, under Harold Wilson, was very pro-nuclear and the established view was that the coal industry had no future and should be run down. The Labour party, in opposition in the early 1960s, had promised support for coal but, in government from 1964-70, reneged on their promises and accelerated the coal industry shutdown.

The campaign reached a climax in 1967-68 with the run-up to the 1967 White Paper on Fuel Policy, and the decision on the proposed AGR nuclear power station near Hartlepool, south of Newcastle. Robens realized that if this station was built, on the very edge of the Durham coalfield, then the coal industry had a bleak future as a supplier of coal to the electricity industry. Roger Williams agreed, writing in *Nuclear Power Decisions*: 'If coal could not compete successfully with nuclear energy there, then its future was bleak indeed. The psychological stake for the mining community was thus enormous'. Feelings amongst miners ran high, as Robens remarked: 'Even to suggest putting a nuclear power station on the coalfield was regarded by the miners as the kiss of death and an act of treachery on the part of the Labour Government'.

Robens justified his campaign by saying that there were half a million people on the coal industry's payroll. He, along with the 1967 Select Committee on Science and Technology, repeatedly called for an independent financial investigation of the economics of nuclear power as he did not believe CEGB claims that nuclear was cheaper than coal. He complained bitterly about nuclear secrecy on costs, writing: 'This conspiracy of silence is a sinister development. Parliament is being denied information of the greatest national importance – information that is essential for seeing whether the vast sums of money being invested by a nationalised industry are being wisely spent'.

However the government had been persuaded, by a 1964 review, that nuclear power in the form of the AGR was the power station of the future. The result, however, was a foregone conclusion: the media and many MPs, such as Tony Benn, were solidly pro-nuclear and there were then few anti-nuclear critics in Britain. One critic was Duncan Burn, who in his book *The Political Economy of Nuclear Energy* (1967) savaged the AGR and Britain's nuclear policy, stating: 'The policy of having a large programme of power stations, originally planned to cost £1,000 million by 1965, was a badly judged use of resources, which was not imposed by what was wrongly called the "energy gap". The UK government's central guidance and control led to a use of lavish resources wastefully'.

However leaders of both main parties in Parliament joined in expressing disapproval of Burn's book. In a later book *Nuclear Power and the Energy Crisis* (1978), Burn reviewed the 1967 debate and wrote that Tony Benn, then Minister of Technology, had concluded that Burn's analysis was 'polemical' and the picture it presented was 'false'. Benn had expressed his 'belief that this country had been given good value for the money it has spent in the field of civil nuclear power' and that 'the position will be seen more clearly in our favour in a few years' time'.

Schumacher's 1967 speech

Fritz Schumacher had long been opposed to nuclear power on ethical grounds. In a 1955 essay *Economics in a Buddhist Country* [32] he called it 'violence against nature' and said that 'Atomic energy for "peaceful purposes" on a scale calculated to replace coal and oil, is a prospect even more appalling than the Atomic or Hydrogen bomb'. He went very public with his views in a speech on 19 October 1967 (the annual Des Voeux Memorial Lecture on 'Clean Air and Future Energy – Economics and Conservation' given to 900 delegates at the National Society for Clean Air conference at Blackpool) which caused a storm. In it he raised doubts about the safety of nuclear reactors, drawing attention to the problem of the radioactive wastes and arguing for watchfulness on the effects of this new technology. He argued: 'Of all the changes introduced by man into the household of nature, large-scale nuclear fission is undoubtedly the most dangerous and profound. As a result, ionizing radiation has become... the greatest threat to man's survival on earth'. As Williams commented: 'Schumacher also offered what was for its time a very unorthodox opinion, to the effect that nuclear stations represented an incredible, incomparable, and unique hazard to human life: worse still, this did not enter any calculations and was never mentioned'.

In his speech Schumacher identified nuclear waste disposal as a key problem:

No place on earth could be shown to be safe from radioactive waste products... The most massive wastes are of course the radioactive reactors themselves... No one discusses the humanly vital point that they have to be left standing where they are, perhaps for thousands of years, an active menace to life silently leaking radioactivity into air, water and soil. No one has considered the number and location of these satanic mills which will relentlessly accumulate in the crowded islands so that after a generation or two there will be no habitation outside the 'sphere of influence' of one or more of them.

Schumacher's remarks were widely reported in the press with a headline of 'Nuclear stations "hazard to life"' in *The Times* the next day. A statement from the Ministry of Power dismissed his remarks as 'so inaccurate that they cannot be regarded as a serious contribution to any discussion of the subject. This may be explained by his having no scientific qualifications in the nuclear field'. The ministry defended the nuclear safety record by saying in *The Times*: 'His suggestions that safety considerations have been disregarded in the development of nuclear power is totally unfounded. Public safety issues are and always have been of the greatest concern to the government'. Peter Williamson, a CEGB delegate at the conference, linked Schumacher's speech to the campaign run by the NCB in favour of coal and against nuclear power, and told *The Times*:

For Dr. Schumacher to suggest that there is a conspiracy of silence to prevent the public from learning of the health and safety aspects of nuclear power suggests to me that he is more remote from the subject than his paper might imply. It is ironic that this attack on the safety of nuclear power should come from a senior officer of the National Coal Board, although maybe it is just part of the anti-nuclear campaign that you may have noticed in the past few months.

As Geoffrey Kirk, his assistant at the NCB, remarked in his book Schumacher on Energy: 'The response to his lecture was immediate and angry. Officialdom was outraged. He was accused of having spoken irresponsibly and of being guilty of special pleading'. There was a public rebuke by Richard Marsh, the Minister of Power, who described his lecture as one of 'the most extraordinary and least profitable contributions to the current debate on nuclear and coal cost'. Schumacher defended his speech in a letter to *The Times* on 25 October 1967, and stressed that economic considerations should not have primacy over environmental ones. He wrote: 'I ventured to suggest that economic considerations must not automatically be accepted as decisive in such a case... there was no need to be in a hurry to change over from conventional power stations to nuclear stations... I am not alone in taking the view that – in the absence of necessity – even a small amount of genetic damage cannot be justified or excused by economic considerations'.

Defeat

With the publication of the 1967 White Paper (which gave an enhanced role to nuclear and a reduced one to coal) together with the approval given to the Hartlepool nuclear station in 1968, Robens realized that the coal industry had suffered a serious defeat. He wrote: 'The effect on the industry's morale was shattering everywhere' and 'there was no concealing the fact that this was a devastating reverse'. His anti-nuclear campaign had been in vain. Both he and Schumacher left the NCB in 1971, and in neither of their biographies is there any mention of nuclear power after 1968. Robens concluded: 'It saddens me to think that the country has suffered because the advice I and my colleagues (notably Schumacher) gave was ignored by the civil servants and the politicians. We were regarded as bloody nuisances at the time'.

Schumacher felt more bitter, as Barbara Wood commented:

The Labour Government believed in cheap oil and in high technology nuclear energy... Wilson's Government, blinded by the benefits of short-term, cheap energy, turned their back on the people; blinded by the seductions of high technology they put their money into nuclear-powered stations and forced the pace of pit closures to accelerate. Life under Labour was far worse for the industry than under the Tories. Fritz felt disgusted and betrayed.

The organic tradition

Schumacher was a supporter of the arguments on the need to conserve natural resources and to respect nature. This conservation position was expressed in 1962 in a paper *Some Problems of Coal Economics*:

Even with renewable primary products, man is not wholly in control of the productive process as he is in a factory but first must fit his actions into the rhythm of the seasons and the often mysterious requirements of organic life. His responsibility cannot be confined to making ends meet and maintaining his man-made assets: he must also 'conserve' the natural conditions which make primary production possible. Thus he has to conserve soil fertility, to conserve forests, to conserve fish populations, and so on.

His use of the words 'organic life' and 'rhythm of the seasons' are, perhaps, the influence of Soil Association that he joined in the early 1950s. According to his daughter, he was strongly influenced by their ideas on organic agriculture, and 'wholeness' and health in the broadest sense: 'His eyes were opened to a whole new way of thinking'. He eventually became their President in 1970. Thus in his 1967 speech he used the language and philosophy of conservation ethics when he said:

In the blind pursuit of immediate monetary gains modern man has not only divorced himself from nature by an excessive and hurtful degree of urbanisation, he has also abandoned the idea of living in harmony with the myriad forms of plant and animal life on which his own survival depends; he has developed chemical substances which are unknown to nature... The religion of economics... promotes an idolatry of rapid change... The burden of proof is placed on those who take the 'ecological viewpoint'.

The publication of *Small is Beautiful* (1973) brought him much publicity for his ethical views. In it his 1967 speech is reprinted in a chapter with the provocative title 'Nuclear Energy – Salvation or Damnation'. He concluded the chapter with a plea for the supremacy of ethical over economic values:

No degree of prosperity could justify the accumulation of large amounts of highly toxic substances which nobody knows how to make 'safe' and which remain an incalculable danger to the whole of creation for historical or even geological ages. To do such a thing is a transgression against life itself, a transgression infinitely more serious than any crime ever perpetrated by man. The idea that civilisation could sustain itself on the basis of such a transgression is an ethical, spiritual and meta-physical monstrosity.

Resurgence and the new environmental message

Although Schumacher was not to receive any press publicity from 1967 to 1973, he took advantage of the new environmental magazines to publicize his views. The first of these magazines was *Resurgence*, started in May 1966. Initially it saw itself as a pacifist journal with anarchist and poetry overtones. However within a year it was publishing articles by Fritz Schumacher, Michael Allaby and John Seymour, to be followed by reprints of articles by US authors Barry Commoner and Paul Ehrlich. Its importance to the environment movement is as a bridge between the old pacifist and conservation tradition and the anarchist and countercultural activists of the 1960s. It was thus able to give the conservation message a new image, one far more anarchist, vegetarian and communal. It was able to make organic farming an essential part of rural communal life, and reha- bilitate the Soil Association from its fringe right-wing position in the mid-1960s to become one of the most respected bodies in the new environmental movement in the 1970s.

John Papworth

Resurgence was started by John Papworth who remained its editor until 1972. He was born in 1920, served in the RAF and trained as an economist at the London School of Economics. He was a life-long pacifist, first joining CND and then the Committee of 100. He was sentenced to one month in prison for his part in the sit-down outside the Soviet embassy on 31 August 1961, and he remained in the Committee of 100 until its end in 1968. As Meredith Veldman, one of the few people to examine his contribution, remarked in her book *Fantasy, the Bomb and the Greening of Britain*: 'Like a number of Committee members, Papworth's protest against the bomb led him to a wide-ranging critique of the society that had produced the bomb. As early as 1961 he began to connect the large-scale tech- nological and economic structures of industrialized society with both the problem of alienation and the threat of energy shortages'.

He wrote regularly in *Peace News* in the early 1960s, being their correspondent in Africa, and developed similar ideas to Schumacher on the threat of large-scale technology and the need for human-scale structures. Papworth believed that the only hope for humanity lay in the creation of small, relatively independent, self- governing communities designed to human, rather than machine, scale, and the reassertion of the primacy of religious values. In the first editorial in *Resurgence* in May 1966 his emphasis was on the search for a solution to large-scale organi- zations, which he described as totalitarian, state-power giants that threatened worldwide war and the squandering of natural resources. He believed that the only alternative was one 'based on love, non-violence, personal dedication and the power of the individual to make his own decisions'.

Fritz Schumacher, with whom Papworth shared many ideas, was a regular contributor to *Resurgence* and joined the editorial advisory group in 1969, becoming an associate editor in 1970. Schumacher's first article in *Resurgence* was in the second issue, where he wrote on intermediate technology and warned on the 'critical situation that may develop during the next twenty years or so on account of the steep and accelerating rise in world energy demands'. He wrote again in the third issue on the industrial society from the Christian point of view, based on R. H. Tawney's ideas.[33] Further articles appeared in no. 7, on the economics of intermediate technology, and no. 11 on Buddhist economics – all themes that were to be used in *Small is Beautiful* in 1973.

Another early ecological writer was Michael Allaby, then assistant editor of the Soil Association's quarterly journal *Mother Earth* (there was also a newsletter called *Span*). In an article entitled 'Politics of Nutrition' in no. 5, Allaby put forward the organic tradition's view of the link between modern farming methods, poor nutrition and degenerative diseases. In the same issue there was an article by David Cooper, the radical psychiatrist, attacking factory farming as a symbol of oppression and violence not only to animals but also to humans.[34] This unself-conscious mixing of the old conservation ideas on health and nutrition with new countercultural ideas on state oppression and mental health was unique. It enabled the old conservation ideas to reach a new audience – the young new left. It gained acceptability or 'hipness' by association, like some dowdy country cousin shown around the city 'scene' by her more fashionable urban relative.

Another writer on rural issues was John Seymour, then a writer of British travel guides and about Gandhi, who had spent time in India. After having a smallholding in Suffolk, he took over a neglected farm in Pembrokeshire from which he and his wife Sally were to write and achieve widespread fame with their book *Self-Sufficiency* (1973). In an article in no. 12, Seymour praised the traditional smallholders and small farmers for their independence, and urged those dissatisfied with centralized society – 'the Monster' – to become a 'modern Homesteader'. Seymour argued that people should withdraw from 'the Monster' and be as self-sufficient as possible: 'He fights the Monster – not by words or with bullets, but by quietly withholding his labour, his custom, and his support. He buys *as little as he can* from the Monster. He contributes to it as little as he can'.

This message of the possibility of escape from urban life was to prove seductive for some in the underground, and the communes movement was to embrace it heartily; their first commune, Selene, had been founded in November 1965. Back to the land was not only 'hip' but also political and ecological.

Resurgence not only published articles by older anarchist writers, such as Herbert Read (b. 1893), Leopold Kohr (b. 1909) and Paul Goodman (b. 1911), but also younger ones in their 20s and 30s (like Dave Cunliffe and Satish Kumar) who were well-known in 'underground' circles. Cunliffe was a radical poet and vegan who had been prosecuted for 'indecency' in 1965, while Satish Kumar was an Indian monk turned 'peace guerilla' who after a two-and-a-half-year journey on foot around the world arrived in London and started the London School of Non-Violence that gained publicity in the underground press in 1969.[35] One writer who moved easily between publications was Geoffrey Ashe, academic and writer on King Arthur and Gandhi. His work on King Arthur, Glastonbury and English mysticism found easy acceptance in the underground press, appearing in *IT* (no. 38, August 1968) and in *Gandalf's Garden* (no. 4), while his work on Gandhi appeared in *Resurgence* (vol. 2 no. 1 and vol. 3, no. 12). From late 1973, with the appointment of Satish Kumar as editor, he had a regular column.

The first explicit articles in *Resurgence* on ecological issues appeared in late 1969 (in vol. 2, no. 10) in a special issue on the 'The Politics Of Pollution'. This contained a six-page editorial, an article by Barry Commoner and a short reprint of an article from the US pacifist magazine *WIN* which had just had a special issue on ecology (August 1969). In the introduction to this article Resurgence remarked:

> This is the first time that an established peace journal has taken this considerable step of relating the question of peace not simply to that of abolishing conscription and war, but to the massive corpus of pressures and policies that is rapidly reducing our planetary habitat to a shambles… Resurgence readers will be aware that we have been pleading for precisely this widening of outlook since our inception over three years ago.

The introduction went on to argue that 'our repudiation of violence must involve a repudiation of industrial and agricultural economies which involve a sustained assault on the habitat as a matter of course, and their replacement by forms which enable men to co-operate with their habitat and with each other'. It is interesting to note that the old ecological word 'habitat' is used, as the expression 'the environment' was not then used in radical circles. The *WIN* article uses the terms 'ecological crisis' and 'ecological consciousness', and mentions the work of Ecology Action, a radical group in California. Commoner, perhaps reflecting his 'environmental biology' background, uses the terms 'environmental crisis' and 'environmental problems' in his article on the dangers of chemical fertilizers in agriculture.[36] The introduction to his article noted his support for the views put forward by previous Resurgence contributors on this topic, including Schumacher.

In the next year there were further articles on ecological issues by Yehudi Menuhin, president of ConSoc from 1969-1970, and by Paul Ehrlich, well-known US eco-doom writer and also to be president of ConSoc from 1971-72. Yehudi Menuhin, who was a supporter of *Resurgence* since its earliest issues, wrote a largely philosophical essay on the relationship of humans and nature, and the meaning of 'conservation'. Paul Ehrlich's article 'Eco-catastrophe!' was reprinted from the US radical magazine *Ramparts*, and was an eco-doom story or 'grim scenario' about the environmental crisis likely to happened in the 1970s, probably based on his 1969 sci-fi novel *Ecocatastrophe*. At the end of the article, he argues in support of his scenario, saying: 'Most of the people who are going to die in the greatest cataclysm in the history of man have already been born', and arguing that population will outstrip food supply leading to inevitable famine. In a forerunner to the notorious computer curves contained in *The Limits of Growth* he says: 'The shape of the population growth curve is one familiar to the biologist... A population grows rapidly in the presence of abundant resources, finally runs out of food or some necessity, and crashes to a low level or extinction'.

The launch of the *Blueprint for Survival* in January 1972 attracted widespread press publicity, and a rebuttal from *The Economist*. An editorial in *Resurgence* (in *March 1972) defended the Blueprint's* analysis but questioned its authoritarian and centralist solutions, saying they were quite incompatible with the existence of small independent communities. The editorial outlined the magazine's difference with *The Ecologist* by saying: 'Our problems are not technical; at root they are moral. We no longer have the kind of communities where people are able to make considered moral choices about their common affairs. That is the historic root of our current crisis'.

This issue of *Resurgence*, with its emphasis on the primacy of moral values, marks its separation from the emerging environmental press, which was to give attention to political action and practical solutions. It was no longer the only alternative magazine giving space to environmental issues. New magazines had started: *Your Environment* in December 1969, *The Ecologist* and *Country Bizarre* in June 1970, and *Undercurrents* in January 1972. Later that year was to come *Seed* in April and *Towards Survival* in June. From 1971 the underground press was giving more space to ecological issues: Joy Farren wrote regularly in *IT* from February 1971, Barry Commoner's work was reprinted in *Frendz* no. 5 (July 1971), and Michael Allaby wrote in *Oz* no. 41 (April 1972). There were also increasing reports on the activities of eco-protest groups such as the Dwarfs (or Dwarves), the FoE bottle campaign and the Liverpool branch of FoE in *Fapto*, *Frendz* and *Ink*.

In its first five years from 1966-1971, *Resurgence* had played a crucial role in modernizing the traditional conservation message. It had nurtured young writers such as Dave Cunliffe, Roger Franklin, Guy Dauncey, Victor Anderson and Keith Hudson, who were all to write on ecological issue for such magazines as *Global Tapestry*, *Peace News*, *Towards Survival*, and *Undercurrents*. Keith Hudson was to become secretary of the Warwickshire branch of ConSoc, a branch that received widespread TV and press coverage in early 1972 for its exposure of toxic waste dumping in the Midlands. He then became editor of *Towards Survival* and a member of the Club of Thirteen, a discussion group based in Coventry, from which the PEOPLE (later renamed the Ecology, then the Green) party was formed. Victor Anderson was to become a leading activist with Reclaim the Streets, which organized anti-car street parties in London in the early 1970s, and with the Young Liberal group Commitment.

Nuclear power coverage

Resurgence as a pacifist magazine was strongly against the nuclear bomb but, like most of the alternative press, it paid little attention to nuclear power. Throughout the 1960s and early 1970s its articles only mention nuclear power twice, in the third and the eleventh issues. In the third issue (Autumn 1966) there was a collection of quotations, compiled by a Dr K. Turkisher, on the problems of atomic waste disposal. Starting with a quotation from Denis Gabor (1963) saying that 'a world power economy based on nuclear fission is very uncomfortable to contemplate', it went on to discuss the problems of radioactive waste disposal in the sea or underground. Finally it concluded with a quotation from the *Bulletin of the Atomic Scientists* (October 1963) which said that unless the waste disposal problem was solved 'it will be impossible to build large atomic power stations on a wide scale'.

What is interesting about this article is how closely Schumacher's speech a year later in October 1967 followed its arguments on the dangers of, and problems posed by, radioactive waste disposal, although he does not use any of the five references given. Schumacher undoubtedly saw this article, as he had an article in the same issue. What is interesting to speculate is whether Schumacher had any part in arranging for this article to be published. And who is Dr. Turkisher: was it one of Schumacher's many pseudonyms?

The next reference to nuclear power was in the eleventh issue (January/February 1968), just after Schumacher's speech to the Clean Air Society's annual conference, which had attracted much press attention. The profile of Schumacher, who has an article on 'Buddhist Economics' in that issue, said:

> A speech he made recently… led to an angry response from nuclear energy officials who betrayed a doubtless well-justified touchiness on the subject of nuclear waste and genetic safety. It even led to a question in the House of Commons, with a minister ponderously agreeing that Dr. Schumacher's speech was 'irresponsible'. He remains, needless to say, unrepentant, as well he might, especially since the conference greeted his speech with a standing ovation.

John Papworth, in his editorial reviewing possible future energy sources, describes 'atomic waste' as 'a material so virulently poisonous as to constitute a deadly genetic hazard should it be leaked into the life cycle'. He bemoaned the near total indifference of even the educated public and described the folly of governments and scientists as 'really past belief'.

There was no further mention for another two years, then there appeared a small box headed 'Stop Nuclear Power Production' in the autumn 1970 issue. It said that 'A meeting to convene an International Association to ban all Nuclear Power Production is to be held in London shortly'. The contact address given was the editorial office of *Resurgence*. Also in that issue in the listings section is mentioned 'Nuclear Power Menace: a brief fact sheet' from Jeffrey Bond with *Resurgence* commenting: 'One man doing his thing on one sheet of paper, result – a concise compendium of the salient aspects of the problem with numerous follow up references'.

Schumacher returned to the possibility of energy shortages in the May/June 1970 issue. He gives nuclear energy a brief mention, questioning whether there would be sufficient uranium for a large-scale programme, but says even if there was, there remains the problem of highly radioactive wastes. 'It is hard to imagine a greater biological threat, not to mention the political danger that someone might use a tiny bit of this terrible substance for purposes not altogether peaceful'.

Chapter 10

Stourport: Citizens against the nuke

The government decided in February 1968 to change its siting policy to allow the construction of nuclear power stations in more built-up areas than previously. This may have pleased conservation groups, like the Nature Conservancy, but the result was to increase opposition to nuclear power stations amongst the proposed host communities. Although there was an uncontroversial acceptance at Heysham in 1969, a proposed station at Stourport (near Kidderminster) was withdrawn before an inquiry after strong local opposition in 1970. There were two public inquiries in 1971 – the first for a decade – over the proposals to build AGRs at Connah's Quay (near Chester) and at Portskewett (near Chepstow). The first lasted 28 days, from 9 March to 6 May, and the inspector, for the first and only time at a public inquiry into nuclear power, refused consent. The second in June lasted only three days; approval was granted but the station was never built.

The campaign against the Stourport nuclear power station in early 1970 illustrated many of the perennial themes of the British nuclear debate: nuclear utopianism by its supporters, arrogance and dismissal of fears by the CEGB, a backlash against hype and the suppression of debate, the founding of a campaign by marginal political elements, lack of support from established conservation groups, and the strength of fears about the hazards of radiation. It was a local campaign, with little national media or conservation group interest or support.

National press coverage

The national press only covered the beginning and the end of this event, and attempted to provide some explanation. The proposals for two AGRs to be built next to two existing coal stations at Stourport-on-Severn, in Worcestershire, was announced in *The Times* on 20 January 1970. There was no further national publicity until nine months later, a few days before the decision was announced to withdraw the proposal on 22 October 1970.

A report in *The Times* on 19 October stated that there had been many objections to the proposal, including those from Worcestershire County Council, the Worcestershire branch of CPRE, the Severn River Authority and the citizens of Worcestershire, who had signed petitions. *The Times* seemed confused as to whether this opposition was more against pylons than the power station by reporting that rural conservationists had recently been saying that power stations are often less of a threat to the countryside than the big pylons which connect them to the national grid. However it admitted that there was opposition to the power station by saying that 'The countryside lobby has also been displaying scepticism to the generating board's stock answer on "thermal pollution"'.

However five days later *The Times* (24 October 1970), reporting on the government's decision to refuse consent, now remarked that this decision was seen as an important environmental victory. In face of a virtually united front from the whole spectrum of local government in the region it said the proposal 'never got off the drawing board'. *The Guardian*, on the same day, reported however that rejection had been based 'mainly on grounds of safety'. Another factor in the proposal's defeat was that Peter Walker, the newly appointed Environment Minister and MP in an adjoining constituency, opposed the scheme. The official grounds for refusing consent were that the site did not meet the government criteria (of February 1968) for building AGRs close to urban areas.

Ian Welsh, in *Mobilizing Modernity* (2000), attributed the withdrawal of the Stourport application to the government's fear of a public inquiry where its siting policy would have been subject to expert examination. Given the absence of any detailed technical report to legitimize the near-urban siting policy, an extensive cross-examination would have been potentially embarrassing and damaging. Welsh argued that 'to have allowed the Stourport application to proceed to the inquiry stage would have been to run the risk of placing expert dissent over the effects of low-level radiation and safety of the AGR firmly in the public eye'.

Local opposition

The proposal for the Stourport nuclear power station sparked much local press coverage and debate. The two local papers, *Kidderminster Times and Stourport News* (*KT&SN*) and *The Kidderminster Shuttle* (*KS*), naturally gave the proposal headline status, and their letters columns were the outlet for an emerging opposition.

The proposal was greeted with initial enthusiasm, with the *KS* (23 January 1970) carrying the headline 'Stourport a "Boom" Town' and quoting Cllr. Stanley Jones, the chairman of local planning committee, as saying 'everyone will get very prosperous from the £100 million station'. This would be because several thousand workers would be employed, and the station would double the ratable value of Stourport. He dismissed any possibility of dangers, saying that it was obvious the government would not allow it to be built if it was unsafe. Stourport Council, he said, was unanimously in favour; in fact it had encouraged the CEGB to build it, having been in negotiation with them for several years. The article noted that this would be the first nuclear power station to be built inland, also the first within an urban area. After this euphoric support from a leading local politician it is intriguing to see, a week after the proposal was announced, that the other local paper, the *KT&SN* (30 January 1970), had the headline on its editorial column entitled 'Nuclear power station opposition expected'. The author commented prophetically or invitingly that 'one can expect the opponents to the scheme to show their hands. In fact there are already indications that some residents of the town are far from keen on the idea… although there has been some talk of Stourport becoming a "boom" town, many people will need more convincing proof that the erection of the £100,000,000 plus nuclear station will indeed bring greater benefits to the majority of the inhabitants'.

The column then went on to discuss how supporters of a proposed bypass of the nearby town of Bewdley were now having doubts, due to fear of modernization of the town. It quoted the views of David Edwards, a local craftsman: 'Why is it that so many people today are striving for something worthwhile to look at – such as representational painting or a Tudor half-timbered cottage?… Because such things were made with the hands, hearts and mind of folk who cared enough for the beauty and above all, the pride of their great English craftsmanship'.

Furthermore Edwards said that 'we are bowing down to the march of the planners and to Big Brother who likes to use their flag of progress as an excuse to hide behind, while they are destroying our countryside further'. Whether the author of this column intended to link expected opposition to the Stourport nuclear power station with the fear of modernization existing in a small rural town is unclear. But doubting questions and criticism were soon forthcoming in the letters column of both papers. In the *KS* of 6 February, James Millington wrote posing questions, while on 13 February a Prof. Bono Publico[37] wrote supporting Millington's doubts. Interestingly Publico raises the issue of nuclear safety, mentioning the Windscale accident of 1957, and stressing the need to study the effects of radiation emissions.

In the *KT&SN*, for the first three weeks of February, there were nine letters printed, only one in favour of the proposed power station. To counter this adverse comment Cllr. Stanley Jones accused his critics (in KS of 20 February 1970) of being 'stirrer-uppers', in an article entitled 'Nuclear Power will bring better life for

Millions'. In it he spoke of the need for more electricity in the West Midlands due to economic growth, and the threat of power shortages if the station was not built. His nuclear euphoria was not shared by all his fellow councillors, with Cllr. A. Hall saying that this 'cannot be substantiated by the facts' and in fact fewer people would be employed in the new nuclear station than at the current coal-fired station (KS 20 March 1970).

CEGB arrogance

At the end of February the CEGB tried to win support with a four-day exhibition with experts available to answer questions. The *KS* (27 February 1970) with its front page headline 'Nuclear Power: The Big Choice' reported on this exhibition. It also mentioned that Cllr. Betty Gazard, one of the few Stourport councillors against, was with her husband Cyril collecting signatures for a mammoth protest petition to be lodged by 14 March at the Ministry of Technology. The article said that the CEGB was dismissing fears of radioactive dangers, and (very interesting for 1970) claiming that it was not a bureaucratic juggernaut. It was quoted as saying: 'We are not a huge powerful, unapproachable authority trampling down all before us in pursuit of some selfish purpose of our own'. Thus the CEGB, even in 1970, was acutely aware of its image and unpopularity, and its reputation for arrogance. However its refusal to acknowledge residents' fears on nuclear safety and the dangers of radioactivity made it impossible for it to shake off its reputation as an unresponsive and aloof organization. Further its inability to provide a convincing answer to the frequently asked question (dating back at least to the first nuclear inquiry at Bradwell in 1956) of why, if nuclear power stations were so safe, were they built in remote areas? And why were there such detailed emergency measures in the event of any release of radioactivity? The change in siting policy in February 1968 to allow stations to be located in built-up areas was accompanied by detailed arrangements for the emergency evacuation of all residents within a radius of about two-thirds of a mile of nuclear reactors, together with curbs on housing development within a radius of two miles. It was these emergency measures that provided the rationale for Worcestershire County Council opposition.

Opposition mounts

In the *KS* of 6 March 1970, the front page story was 'Nuclear Power Station: County Probe'. This reported that a sub-committee of the county planning committee was to look into the whole matter, particularly the question of radioactive hazards both in the air and in the water discharged into the river Severn. This county action was perhaps in response to increasing local opposition, for Cllr. Betty

Gazard, in a letter to *KS* of 20 February 1970, made an appeal to people to write in to the Ministry of Technology and object to the proposed station, and she said petitions were being organized. This petition was a success for the *KS* (27 March 1970) in its front page story '1,700 Sign N-Protest' reported that Cyril Gazard had sent a petition bearing 1,700 names to the Ministry of Technology. The *KS* reported that this petition had a list of ten objections to the proposed station, and it remarked on the wide area from which signatures had been collected.

The first calls for a public inquiry came in mid-March. In a front page story, 'Nuclear Power "An Inquiry is Essential"', the *KS* reported (13 March 1970) that the county council had lodged a formal objection, and that Norman James, a local councillor, said that a public inquiry was 'absolutely essential'. The proposal was causing splits in the Stourport Council. In a letter to the KS of 20 March, Cllr. A. Hall complained about Cllr. Stanley Jones 'attempting to steamroller this matter through and using dictatorial methods to silence anyone who opposes him'. He again made this complaint at a public meeting in June, reported in *KT&SN* of 5 June 1970, by saying about Cllr. Jones: 'He has consistently done his best in the past to stop questions being asked in the council chamber'.

By the beginning of April, the *KS* headline (3 April 1970) was 'Second Thoughts? Opposition to N-Plant Mounts at Stourport'. The report said that a powerful group of councillors might persuade Stourport council to change its mind. The recent Maud-Ratcliffe report on local government financial reform meant that the town would now not be able to keep the rate revenue from the nuclear station. It also mentioned that Mrs. Gazard now had over 2,000 signatories on her petition (the final count by late April was 2,200). Local MPs were now coming out against the proposal. Peter Walker, a neighbouring MP, was reported as having joined 'the fight' against the power station, while the local MP, Sir Tatton Brinton, had demanded a full public inquiry so that objections could be heard. He asked why, if the nuclear station was completely safe, were there restrictions on the number of people who could live within a two-mile radius?

County council report

The most powerful blow against the proposal was the publication of the report by the planning committee of Worcestershire County Council in late April. In it the committee recommended that at the next council meeting on 4 May, the council lodge an official objection and demand a public inquiry into the proposal. This they duly did and the committee report made headline news in both local papers. The *KT&SN* (24 April 1970) had the headline 'Hundred evacuated if emergency occurs' while the *KS* more soberly had '7 point objection to Nuclear Station'.

The seven points of objection as reported in the papers were:

- The proposal would restrict future residential development (due to emergency arrangements).
- Visual impact of power station with its massive cooling towers.
- Impact on river flows due to the massive need for cooling water.
- A degree of risk posed to nearby residential areas.
- Loss of amenity and lower house values to those living near the power station.
- Burden of maintaining emergency arrangements.
- Stourport was already too crowded.

Minutes of the county planning committee revealed that it was the restrictions imposed by the emergency arrangements that were the deciding factor. The minutes of the special sub-committee meeting of 3 April stated: 'In the light of restrictions upon development... we recommend the Committee to oppose the proposal on general planning grounds because of the restrictions which would result upon housing development in Stourport... We also suggest that the Committee consider an amenity objection based on the massive structural dominance of the proposed reactor and turbine house buildings together with the cooling towers'.

Radiation dangers

The recommendations of the special sub-committee were duly accepted by the planning committee on 22 April and they also identified other concerns including danger to health from radiation. The sub-committee had written for advice to a number of organizations, both in this country and abroad. They decided to ask Professor D. C. Leslie and Dr. J. Shaw of Queen Mary College, University of London, to give an independent evaluation of the CEGB proposal. Their report was reviewed by Professor P. W. Spiers of the University of Leeds, who gave particular emphasis to their views on the opinions of Gofman and Tamplin who had made controversial statements on the dangers of radioactivity.

Leslie and Shaw produced a 3-volume report that gives an interesting view of expert opinion on nuclear power in 1970. Broadly they were in favour, with some technical reservations, and they much preferred nuclear to coal stations. They stated:

Provided that the Council is satisfied about the safety aspects, the AGR would appear to be preferable in every way to the coal-fired station. The chief advantages of the nuclear station are:

1. It is smaller.

2. There is no coal dump.

3. There is no chimney of any size.

4. No combustion products are discharged.

5. Once construction has been completed, the traffic to and from the nuclear station will be very much less...

It is interesting to note that there is a discharge of radioactivity from a coal-fired station: depending on the type of coal used, this can be comparable to the discharge from a nuclear station. When considering the potential health hazards of a nuclear power station, it should be remembered that the discharge of combustion products from a coal-fired station represent an actual and unavoidable health hazard.

In their recommendations they stated: 'We consider that, provided there are no overriding objections to it, the proposal to site an AGR power station at Stourport is in the national interest. The demand for electricity is increasing, and we must hope that the rate of growth of demand will itself increase as the economy recovers... we feel that, provided the Council is satisfied about the radiation safety aspects of the proposal, a nuclear station is much to be preferred to one fired by a conventional fuel such as coal'. They considered three possible objections:

1.   Effects of steam plume.

2.   Excessive demands on, and damage to, the river.

3.   Radioactive hazards.

In their discussion of radioactive hazards they discussed the consequences of a possible reduction in radiation doses to the general public. They mentioned the work of John Gofman and Arthur Tamplin in the USA on the need for a reduction in permitted doses. They stated that 'Gofman and Tamplin are responsible people' and 'should not to be confused with Professor Sternglass, who recently got a lot of publicity for his views that bomb tests had killed hundreds of thousands of children. Our colleagues are unanimous that the statistics will not bear the interpretations which Sternglass is trying to put on them'.

They posed the question of what would the CEGB do if standards were tightened. Their report concluded by saying: 'While we think that the proposal is basically advantageous, we have a number of reservations about it. Those are mainly technical and we advise that the Council should attempt to settle them by negotiation... before going to a Public Inquiry'.

Professor Spiers was even more pro-nuclear. In a letter to the county health department he wrote:

1. The site is near a populated area as distinct from the sites so far used.

2. This practice is almost certain to increase as more reactors are built in this small

highly industrialized country… proposals for Stourport will have been very carefully drawn up by experts to ensure that radiation levels and effluent discharges are well within what are internationally accepted levels.

On the question of a possible accident, he wrote: 'I believe Windscale-type accidents are unlikely or impossible with the present closed-circuit type of reactor, but I imagine some kind of accidental escape is conceivable and I imagine the Council to be concerned about this, particularly in view of the recently renewed press and TV discussions on radiation dangers'.[38]

Gofman and Tamplin's assertions that the existing maximum permissible levels should be lowered, was too much for Professor Spiers. In his brief review of the report by Leslie and Shaw, he stated that Gofman and Tamplin were combining the most pessimistic risk estimates of the ICRP (International Commission on Radiological Protection). Furthermore Spiers attacked their assertions that the ICRP levels should be reduced by a factor of ten to a population dose level of 17 millirems by saying that it 'cannot be very compelling in the light of the variations in the natural dose level – the difference in the means for the population of Aberdeen and Anglesey already differ by twice this amount'. Thus the 'nuclear experts' could find no grounds to oppose the CEGB proposal on radioactivity grounds, despite public fears.

Nuclear siting policy

One of the key questions was whether the proposed nuclear station at Stourport would meet the government's new policy on nuclear siting. A short report commissioned by the sub-committee from the consultants, Associated Nuclear Services, reported: 'It is doubtful whether the choice of site is a reasonable interpretation of the Government's policy statement of February 1968 in view of the intensive development close to the proposed site'.

Leslie and Shaw however were of the opinion that it was not feasible to oppose the new national policy to allow nuclear power stations to be sited much nearer to built-up areas than had been the practice before 1968. The meeting of the planning committee on 22 April 1970 then considered the question of whether the Stourport site is so near a built-up area that it presented an unacceptable hazard. Their minutes reported that 'Dr. Gronow (an expert from the Ministry of Technology) told the Sub-Committee off the cuff that Stourport appeared to be well below Heysham (a comparable urban site in Lancashire, but not yet built and tested) as regards the elements of risk by inhalation'.

It thus appears that the committee did not have evidence to oppose the station either on siting grounds or risk grounds. Despite this W. J. Balderstone, the chairman of the planning sub-committee, presented his own paper entitled *Engineering Aspects of a Nuclear Power Station at Stourport* to the meeting. The minutes stated that this paper 'was intended to show the many points at which leaks of radioactive material *could* occur'. In a prophetic sentence, which was reprinted in the *KT&NS* of 24 April, Balderstone warned: 'Human error, carelessness and lazy indifference are liable to occur in the management of any plant, however well designed, especially perhaps at night when the rest of the world is asleep'.[39] He further warned: 'Already insurance companies have notified householders in the immediate neighbourhood that nuclear risks are excluded. The emergency arrangements, however well-presented, must be a source of anxiety and are likely to dampen the ready sale of houses. The fear of a much greater danger than really exists will prey upon the minds of many people'. This issue of risk was at the heart of the debate on the proposed station. Associated Nuclear Services commented: 'Behind the Government policy statement there is a broader principle that in siting the nuclear power station there should be an appropriate balance between risk and benefit, and in this case it seems most unlikely that the additional benefits in the form of reduced costs will compensate for the additional risk to the local community'. The planning committee seemed to have adopted the ANS view, for it stated:

> The Council as planning authority may well decide that the applicants should be put
> to the test to prove that the risk does not exist or is justified in the national interest
> by the counterbalancing gains. If so, a suitable wording for this ground of opposi-
> tion might be:
> 'The proposed site so near to built-up areas presents a degree of risk which has not
> yet been justified'.

This demand that proof be provided that 'risk does not exist' was to become a central theme of anti-nuclear protesters in the 1970s. In contrast the argument that nuclear stations were 'justified in the national interest' became a major plank of governments' attempts to push forward their nuclear policy.

Stourport: the end

With the publication of the planning committee's report in late April, and the proposal's negative portrayal in headline news in the local papers, the proposal was effectively dead. The local campaign nevertheless intensified. In early June there was the founding of the Anti-Powerport Station (APS) with a committee composed of many councillors and with sixty members. At a public meeting organized by the APS, reported in the *KT&NS* of 5 June, there were calls for a register of objectors, and 'militant anti-nuclear power station rallies'. There was further talk of raising money to employ a barrister to state their case at a public inquiry. Cllr. Eric Higgs, the secretary of the APS committee, was reported as saying: 'We are objecting because we do not want this in Stourport. There will always be an element of fear, and we don't want to live with it'. Similarly Cllr. Betty Gazard, a prime mover behind the campaign, said: 'None of us know how much we shall receive, or if we shall receive any, or if it will do us any harm'.

Schumacher's idea that nuclear power was a 'transgression against life itself' made three years earlier chimed in with residents' concerns about the effect of radiation on future generations. As a local headmaster, Mr. Baylis, said: 'There are some people in Stourport who have not been consulted – the children. If we hand this down to the children of Stourport as a heritage, we shall have done them a great deal of harm'. Interestingly there are no further reports after this date in the *KT&NS* of the nuclear proposal or of the APS, only a few letters mainly about conflicts over the handling of this issue in the Stourport council. The official announcement in late October that the proposal was being withdrawn was not mentioned. From June onwards it was a dead issue to the local media.

Portskewett

The campaign at Stourport against their proposed AGR had a knock-on effect at the next site proposed by the CEGB for an AGR. This was at Portskewett (located south of Chepstow on the Severn estuary in what was then Monmouthshire) just fifty miles down river from Stourport. The concerns expressed and the associated public campaign at Stourport, together with the actions of the Worcester County Council, were common knowledge in Portskewett. Whilst Monmouth County Council and the rural district councils supported the application, Portskewett Parish Council strongly opposed it. According to the public inquiry transcripts the county council were accused of having delivered the whole project to the CEGB 'on a platter', and the public of Portskewett had 'lost all confidence' in their county officials because of 'their excessive enthusiasm... in projecting the merits of nuclear power' and their 'subservience' to the CEGB.

The concerns of Sternglass, Gofman and Tamplin over low-level radiation were raised once more, and Portskewett Parish Council urged that Sternglass' work be checked by the government. Particularly concerned about exposure to radiation were the local women's group, the Portskewett Young Wives. The CEGB however portrayed these concerns as the product of press reports (such as that on Sternglass that appeared in *The Observer* on 16 May 1971) of discredited scientific work, relating to a different type of reactor, the American LWR rather than the proposed AGR. However public concerns over radiation dangers found limited reflection at the Portskewett public inquiry. The main concern of government and the CEGB was over challenges to its urban siting policy, which lacked any (published) technical basis. Their position at the Portskewett inquiry was however strengthened by the precedent of the granting of a site licence at Heysham in 1969, where the policy had gone unchallenged. The Portskewett public inquiry lasted just three days in June 1971, and the application was approved, but the CEGB never built an AGR there – most probably because by the early 1970s it became aware of serious design defects with the AGR.[40]

Chapter 11

Counterculture and alternative press

The 1960s were a time of much social change and political protest which took the form of various social movements which are termed the 'counterculture' or, to its participants the 'underground' or 'alternative society'. This counterculture, while novel in style and appearance, was however an outgrowth of long-standing social groups and political conflicts in post-war society. As sociologists Braunstein and Doyle commented in their book *Imagine Nation*:

> The Sixties counterculture in the United States didn't come out of nowhere: it appeared gradually as a ripening of popular discontent over America's shrill postwar triumphalism. It was a fruit that had been assiduously cultivated throughout the 1950s in the many scattered patches of bohemia across the land, and across the Atlantic... vigorous denunciation of cold war militarism, anticommunist demagoguery, racial segregation, social regimentation, and rampant, near-orgiastic consumerism articulated most of the themes that adherents of the 1960s counterculture would echo and amplify.

The counterculture initiated few ideas. Mostly it just adopted mainstream ideas, in a very idiosyncratic way, which were then fed back and sometimes slowly adopted (if compatible with a middle-class lifestyle). The same with environmental ideas – mainstream concerns, articulated by the press and TV, percolated through to the counterculture, and any 'hippie' for instance reading *The Guardian* would soon know about pollution and the threat of ecological catastrophe. The response however would be different from that of their parents – a change of diet to wholefoods rather than letters of protest or joining the Conservation Society.

Counterculture

The term was popularized by Theodore Roszak in his bestseller *The Making of a Counter Culture* (1968) and has since become the term to describe sixties cultural radicalism. Its meaning and significance has been much debated in the last decade, with older definitions of it as a social movement with positive political and sociocultural aims being challenged by more recent research. Typical of the older view is Elizabeth Nelson who remarked in her book *The British CounterCulture* that the counterculture was 'an elusive phenomenon' but that it can be considered as 'a social movement with explicit and fundamental criticisms of modern capitalism, which offered an alternative vision… of a millenarian type'. After reviewing many authors on the counterculture she wrote that most observers were agreed that the term 'counterculture', like the 'underground', which is often used synonymously, was almost impossible to define. She considered it however as a struggle against the 'culture of the powerful' with the hope that the power of dominant culture could be broken.

More recent research by Braunstein and Doyle refuted these ideas of purpose and direction:

> The term 'counterculture' falsely reifies what should never properly be constructed as a social movement. It was an inherently unstable collection of attitudes, tendencies, postures, gestures, 'lifestyles', ideals, visions, hedonistic pleasures, moralisms, negations, and affirmations. These roles were played by people who defined themselves first by what they were not, and then, only after having cleared the essential ground of identity, began to conceive anew what they were. What they were was what they might become – more a process than a product, and thus more a direction or a motion than a movement.

The problem with interpretation of the counterculture is that its participants, people labelled 'hippies', were deliberately anti-intellectual and preferred music to books as their means of communication. Feelings and emotions were more important than rationality and logic, they disdained the scientific worldview, and to them inconsistency in beliefs was not a problem. New Age mystical beliefs happily coexisted with love of technical gadgets – the stereo was as important as macrobiotic brown rice. As Braunstein and Doyle commented the counterculture revelled in 'metaphors, unresolved contradictions, conscious ruptures of logic and reason; it was expressly anti-linear, anti-teleological, rooted in the present, disdainful of thought processes that were circumscribed by causation and consequence'.

There is much dispute about the legacy of the 1960s counterculture. Some dismiss it as a failure. However George McKay in his book *Senseless Acts of Beauty* claimed that the utopian project of the sixties is still with us – in fact never really went away – and illustrated his book with an account of politically radical countercultures since the 1970s, or what he called 'cultures of resistance'. He vividly argued that the counterculture of the 1960s cannot be simply dismissed via a narrative of failure, but has spawned generations of activists and cultures of resistance to the present day.

Hippies

The term 'hippies' was used first by the news media about the young people flocking to the Haight-Ashbury area of San Francisco in 1967, perhaps derived from the word 'hip'.[41] For a while it was fashionable and adopted as an ideological charade by some, but this meaning was dropped by 1968-9, after which it was used as an adjective for a look, a fashion, an attitude, or a lifestyle. People in the underground disliked the term 'hippie' preferring to call themselves 'heads' or 'freaks', hence phrases such as the eco-freak press.

Arthur Marwick, in his monumental book *The Sixties*, made the point that the number of people fully involved in the underground was a very small part of the total population, for the USA probably about 0.1% of the population. Thus only a tiny minority could be described as full-time hippies, though there were many times more 'summer and weekend hippies' who were full-time students or had regular jobs but adopted some aspect of the hippy lifestyle (the clothes, the hair or the drugs). Marwick emphasized that there was no counterculture uncontaminated with mainstream culture, and thus there was no impermeable barrier between the two; escape from the 'straight' world to some hippie enclave, such as the rural commune, was an impossible dream.

Technology attitudes

The mid-1960s was a highly optimistic period not only for the counterculture but also for most of society (apart from the poor). This was due to the full-employment prosperity of the era and the prevalent utopian ideas about the imminence of a 'post-scarcity society' in which the need to work for a living would be radically diminished, if not eliminated altogether. This lasted about three years from 1966 to 1969. As Tom McGrath, the founder of *IT* and an ex-editor of *Peace News*, wrote in *IT* no. 10 (March 1967):

> The new movement is essentially optimistic. It has a happy view of man and his potential, based mainly on his creativity. This is a post-existential movement, bringing an end to years of tough and painful despair. That optimism has been reborn in the face of the H-bomb, Vietnam, poverty, hunger, etc., is so surprising it

is almost a miracle. Again the optimism is tinged with commonsense... The big world problems that concerned CND and the Committee of 100 at their zenith have not been forgotten. The new approach is to make positive changes wherever you are, right in front of your nose. The weapons are love and creativity – wild new clothes, fashions, strange new music sounds.

Thus the underground press was initially optimistic about the possibilities of utopian change through the use of technology. Braunstein and Doyle commented: 'The end of work would herald the onset of a vast leisure society in which human pursuits, liberated from the drudgery of alienating, soul-slaying labour, might be redirected to self-actualization involving the cultivation of each individual's creative talents'. This idea of the coming of a post-scarcity society, through the use of automated machinery, influenced the idea that the counterculture would transcend capitalism rather than destroy it. Elizabeth Nelson summed up the attitude of the underground press to technology when she wrote: 'Technology-as-liberator was a common theme in the underground press, notwithstanding that branch of the counter-culture which sought a return to the idealised pre-industrial revolution golden age'. She quoted Richard Neville, the founder of *Oz*, as saying in his book *Playpower* that 'people would be paid not to work because work had become obsolete through the correct application of technology'.[42] This theme of a bountiful and liberating technology was expressed in *Frendz* no. 6 (July 1971): 'Technology's correct use was an essential element in the realization of the full potential of play. Those who feared automation... also feared leisure... it was necessary to get people out of the factories and get the machines in'.

Automation and science was seen as the key to a liberated society. In *Oz* no. 9 (February 1968) there was a vision for an alternative society based on Digger 'love communes' with automation of the means of production. Nelson described this as arguably the most significant issue of *Oz* to be published, containing quite a comprehensive and seriously argued 'blueprint' for the countercultural society. The article called for all readers to create Digger 'love communes' each with its own land, houses and means of production for primary as well as secondary goods. The means of production would be automated; scientific and technological research would be conducted in order to 'easily flood the market with cut-price and free goods'. This was implicitly scientific anarchism – work would be abolished as drudgery would be automated. Thus contrary to many beliefs, the underground press cannot be considered to be anti-science and technology. It was more interested in science fiction than mysticism, in high-technology sex (the contraceptive pill), drugs (LSD) and music (Pink Floyd) than low-technology wholefoods, country crafts and farming work.

As Nelson remarked, the three years 1966-69 were the 'great days' of the English counterculture, 'a period of optimism and confidence, even of the conviction that the counter-culture's millenarian vision could be realized'. This optimism then gave way to pessimism that the 'revolution' would not arrive, that May 1968 was a transient phenomenon. There was then less emphasis on political street fighting and more on the creation of alternatives. Marx was increasingly replaced by mysticism, mass action by individual living, and the urban 'pad' by the rural commune.

Alternative technology

One outgrowth of this swing towards practical alternatives was a new and critical attitude to technology. There was a desire for a new kind of technology – variously labelled appropriate technology, alternative technology or just simply AT – that could be used to create an alternative society. The AT movement was influenced by new left politics with its critical re-evaluation of long-standing assumptions about the relationship between nature, technology, and society. In particular AT rejected the then current notion of the forthcoming technology-based utopia or 'post-scarcity economy' used by social theorists such as Murray Bookchin, and by some hippies to justify their existence. The AT movement, influenced by contemporary environmental concerns over pollution and over-population, instead argued the need to abandon the bourgeois quest for more consumer goods. Natural resources should be redirected towards establishing social equity and ecological harmony instead of consumerism and waste – concepts we would now call sustainable consumption.

The AT movement, combining new left and environmental critiques of society, believed that social and environmental problems stemmed not from lack of resources but from a misguided waste of the 'technology of abundance'. Technology used morally and ecologically could create a utopian future. Those sections of the counterculture concerned with this quest were a diverse group with a wide variety of perspectives, often pursuing opposed or mutually exclusive projects. As Andrew Kirk pointed out, AT enthusiasts embraced the 'seemingly contradictory notion that the antimodernist desire to return to a simpler time when humans were more closely tied to nature could be achieved through technological progress'.

The relationship between the counterculture, technology, and the environment was complex, and in particular the relationship between the counterculture and technology was always one of fundamental ambivalence. Witold Rybczynski, an AT pioneer and later historian of the movement, believed that the counterculture, rather than rejecting technology, was immensely attracted to it. This fact was shrewdly used by Stewart Brand in his vastly popular countercultural book *The Whole Earth Catalog*, first published in 1968 and with many editions thereafter. What were being rejected were the political and social institutions of technology, what Roszak called the 'technocracy'. The American love affair with gadgets, particularly for self-sufficiency and freedom, continued unabated.

The underground press

The underground press fostered the growth of the counterculture by creating its identity, disseminating knowledge, and validating shared understanding. It is the most tangible legacy of the counterculture. In the UK there were (in the early 1970s) nearly 500 papers, covering a wide variety of issues, ranging from nationally distributed magazines with circulations in the tens of thousands, to small local papers with circulations of a few hundred.[43] It has been extensively documented and microfilmed by Harvester Press, and many books have been written on it covering both British and American titles. One of the first was by John Spiers (1974) who listed 83 magazines (but not *Oz*). Of these I consider seven to be concerned with early ecological issues – mysticism/nature, wholefoods, country crafts, rural communes and pollution. They were *Communes, Country Bizarre, Dwarf News, Gandalf's Garden, Global Tapestry, The Snail* and *Torc*, and I term them the rural underground press. Apart from the three big circulation papers (*Oz, IT,* and *Friends* all based in London), the others were 'community' papers or 'specialist' for such groups as social workers, librarians, gays, women etc.

There are few circulation figures for the underground press. The largest, *Oz*, claimed sales of 27,000 for issue no. 18 (September 1967), 40,000 and a print run of 50,000 for no. 29 (July 1970) and 50,000 for no. 32 (January 1971). *IT*, its main rival, claimed sales of over 40,000 in February 1969 with perhaps 150,000 reading it. Allowing for some hype, one can definitely conclude that sales were in tens of thousands. In contrast, although there are no figures, sales of the rural underground press were most probably in the low thousands. For in the 1970s the leading British environmental magazine *The Ecologist* had a circulation of about 3,000 and the leading eco-freak magazine *Undercurrents* had a circulation of about 6,000.

There is, as yet, no index to the underground press to gauge coverage of issues. Cursory reading indicates that the contents fulfil the cliché of 'sex, drugs and rock and roll'. An index to the first twenty-two issues of *IT* (1966-67) reveals that out of 200 articles listed there is one on macrobiotic food and three on 'UFOs and other galactic issues'. Nelson looked at the contents of *IT*, *Friends* and *Oz* on launch, after six and twelve months, and the final issue. The only mention of ecology is in *Friends* no. 19 where there is a critical article entitled 'Ecology

Sucks', which is a US reprint, together with a reprint from the *Whole Earth Catalog*. Thus I conclude, both from numbers of magazines, circulation figures (scanty as they are) and (incomplete) content analysis, that ecological issues were only of interest to about 5-10% of underground press readers.

Rural escapism

Despite there being few ecological articles in the underground press, some observers believed that there was a search for a new lifestyle based on ecological ideas. Michael Allaby, an early environmental journalist, observed in his book *The Eco-Activists* that: 'The "underground" in particular and the whole ecological movement, is based on a new awareness of and care for the world we share with one another and with other life forms. It is but a short step from here to concern about the ways in which the land is treated in order to produce our food'.

He stated grandiosely that the 'underground people' sought to establish a way of life that respected the environment and that they were experimenting and striving towards a new culture and the foundations of a new society, 'a society in which men can live in harmony with nature'. However this vision of the human race living in harmony with nature was shared at that time by few in the underground. Allaby, then editor of the Soil Association journal *Span*, was obviously keen to find young supporters for organic farming. Like many others with a cause he looked to the underground as a source of support. Thus he was prepared to overlook the mystic ravings on the Earth Mother, the Moon Maiden and the Sky God by his interviewees in discussions on rural life and agriculture.

The rural underground press started with *Gandalf's Garden*, which ran for only six issues from May 1968 to November 1969, followed by *Country Bizarre* (eleven issues from summer 1970 to winter 1972), and *The Snail* (started March 1971). There were also more specialized magazines such as *Communes*, *Torc* (on Glastonbury and British mysticism), *Dwarf News* (freaky anarchist) and *Global Tapestry* (pacifist, vegan and anarchist). Also influential, but on the borders of this rural underground, were *Resurgence* and *Peace News*.

This press can be seen as an offshoot of the Romantic tradition in its questioning of authority, reason and science and its support for anarchism, mysticism and nature. Unlike the majority of the underground press, such as *OZ*, *IT* and *Friends*, it rejected belief in technological progress and looked to the past rather than the future. It looked for inspiration to the hobbits of *The Lord of the Rings* rather than to the global village of Marshall Mcluhan and the science fiction of Moorcroft. This search for roots in the countryside has a long tradition in Britain. In their book *Contested Natures* (1998) the sociologists Phil Macnaghten and John Urry commented that the pioneer Victorian preservationists sought to restore links with the past through the preservation of 'relics' and national monuments so as to 'reinvigorate a spiritual relationship to place and country'. Furthermore they highlighted nature as the 'unspoilt' other as embodied in the relics, customs and mystery of the English countryside. This rural nostalgia was continued by most 1970s environmental groups which rejected many of the values of contemporary society and based a great deal of their activities on particular aspects of British culture and tradition, such as preservation of the countryside.

An important practical feature of the underground press was the listings column, with addresses where readers could get advice, help or more information on a topic. These were compiled by the editors, and enabled them to list organizations that were outside the counterculture (and often strongly disapproving of it) but whose knowledge or skills the editors felt their readers should know about. For instance *Resurgence* had a column (in the late 1960s) originally called 'Publications' and later 'The Movement' in which it listed the Anti-Concorde Project, The Conservation Society, the Henry Doubleday Research Association, and Soil Association News alongside pacifist, anarchist and countercultural groups. The juxtaposition of groups was often bizarre; the deeply traditional organic farming group, the Soil Association, was listed next to the Selene Community, an anarchist and free-sexuality commune. The impression given to readers was that organizations such as the Soil Association were as much a part of counterculture as communes or pacifist papers. This association was not however always well received. The Soil Association complained about being listed as an organization that 'shared some aspects of *Resurgence* concerns' and in the next issue there was an apology from *Resurgence* saying: 'We would like to make clear that the Soil Association is in no way connected with *Resurgence* or its policies'. Nevertheless the Soil Association continued to be listed in the Publications column, as it was in many other alternative magazines. In this way conservation organizations were incorporated (most probably deeply unwillingly) into the counterculture. Ecological concerns, such as organic farming, were thus made as valid as communes or anti-Vietnam war protest. In the early 1970s the interests of the rural underground press were to merge with the concerns of the 'straight' environmental press such as *The Ecologist, Towards Survival* and *Soil Association Review*. This was to find expression is such 'eco-freak' magazines as *Aether, Seed, Undercurrents*, and *Whole Earth*, which mixed rural nostalgia with contemporary ecological concerns.

The rural underground press

One of the key characteristics of the rural underground press was its rural location – that is, outside London and other large cities – and its emphasis on rural and countryside issues, albeit seen from a countercultural perspective. Its main interests were, at first, mysticism and eastern religions together with macrobiotics and wholefoods. This mysticism found expression in a revival of interest in British mythology, country crafts and customs and traditional agriculture. The interest in traditional agricultural methods such as organic farming was encouraged, even required, by rural communes seeking self-sufficiency.

Gandalf's Garden

Gandalf's Garden is one of the underground papers most cited by academics. It has had an influence out of all proportion to its circulation, and has been responsible for the myth of the underground's preoccupation with the mystical, particularly *The Lord of the Rings* which has been mentioned by numerous academics and in particular by Meredith Veldman who devotes a large section of her book *Fantasy, the Bomb and the Greening of Britain* to Tolkien (but does not mention *Gandalf's Garden*). Frank Musgrove, an academic, even wrote in his book *Ecstasy and Holiness* (1974) of the underground press, Tolkien and *Gandalf's Garden* as being synonymous: 'On the far outer fringes are relatively inactive sympathizers who join encounter groups and read the underground press, Tolkien's *The Lord of the Rings*, R. D. Laing, *Gandalf's Garden*, and books about the open family. They may be vegetarian, eat health-foods and practice yoga'.

This is highly unlikely as *Gandalf's Garden* was only published for six issues between May 1968 and November 1969. As a magazine it has received critical acclaim from underground press researchers. John Spiers wrote in his book *The Underground and Alternative Press in Britain*: 'The first of the mystical magazines in the underground, *Gandalf's Garden* was a terrific pioneer effort which greatly influenced the rest of the underground/alternative press in the late sixties, particularly in its approach to graphics and its coverage of esoteric religions'. The magazine was founded by Muz Murray, who ran a shop of the same name in World's End, Chelsea which ran classes on eastern religions. After he left England, regional centres were set up around the country called Friends of the Garden. Kenneth Leech, in his book *Youthqake* (1973), wrote that *Gandalf's Garden* 'became a vehicle for the expression of the early London mystical scene. The influence of Tolkien's *Lord of the Rings* was evident in its choice of hero. Gandalf the White Wizard was seen as the "mythological hero of the age"'.

Similarly Nigel Fountain in his book *Underground* gushed that *Gandalf's Garden* 'distilled the essence of the time's peace, love, mysticism and delusion' and that it 'took as its theoretical base – if theoretical is the right word – J.R.R. Tolkien's *Lord of the Rings*... During its short life *Gandalf's Garden* evolved a style which made *It*, *Oz*, and others that were to come later models of detachment and worldliness… It was *Gandalf's Garden*'s task to carry, assisted by divine and hobbit hands, the purest expression of the hippy philosophy across the city, and country. It did, on occasions, meet with approval, but largely, within the underground with incredulity'.

Richard Neville, then editor of *Oz*, was dismissive, describing it in his book *Playpower* (1971) as a 'mystical mish-mash of madness and megalomania' but its influence, like that of Tolkien's, has grown over the years. *Gandalf's Garden* was not only concerned with mystical issues, but also gave space to pacificism, communes and vegetarianism. It carried articles by Tony Kelly on the Selene Commune in no. 3 (see below on Communes) and Dave Cunliffe, the radical poet and vegan, in no. 5 (he was an early contributor to *Resurgence*).

A successor to *Gandalf's Garden* was the small magazine *Torc-Glastonbury Voices* that ran for eight issues from September 1971 to summer 1973. It not only concerned itself with the mysticism of Glastonbury but also 'alternative lifestyles'. Thus it had articles on the origin of Glastonbury Tor, the summer solstice and wholefood cooking, but its main emphasis was on the spiritual power within these islands, a favourite theme of *Gandalf's Garden*.[44] Spiers wrote that Torc combined the mystical with the organic, actively pressing for everyday alternative living:

> To demonstrate that there can be an alternative life-style: an alternative economy based on self-sufficiency and mutual aid; an alternative politics brought down to human-size for human needs whose only authority rests with those having the right intelligence or the right skills who might happen to be on hand at any given situation… The Alternative isn't opting out, and sitting around all day; the Alternative is both hands and feet in the Earth working where all things must begin.

Country Bizarre

Country Bizarre can be considered the first eco-freak magazine. It was widely admired, and Spiers wrote glowingly: '*Country Bizarre*, one of the most deeply satisfying, elegant and beautiful publications of the alternative society, speaks for the gentle revolution and is unique amongst the underground/alternative press in Britain… Produced with care and imaginatively illustrated'. The first of eleven issues was produced from Rochester, Kent, in the summer of 1970 and it ran until winter 1972 (all issues are reprinted in the book *The Complete Country Bizarre* (1976) by its editors Andy Pittaway and Bernard Scofield). It was deeply into country crafts, such as weaving, spinning, and thatching, besides local poetry and 'the nature side of things'.[45]

The first editorial gave the magazine's philosophy, as one steeped in the long-standing romantic protest against technology and industrialization. It began: 'In 1750 saw the beginning of the industrial revolution. A vast uncontrollable upheaval which in its time, changed the life, character and destiny of most of the English people. The general harmony of old and traditional habits was destroyed'. The only solution was to reject the modern economy and go back to the land:

> Technology, in its crazy race, outstrips our comprehension and man seems unstop-pable in his mad career. Industries for industry's sake. An economy run on mass car production and its subsidiaries... No one is going to throw this mighty industrial hen off her nest too easy; she is well bedded down. Our only way is to deprive her of her meal, which can only come about by rejecting the paraphernalia of her goods and reviving the pre-industrial agricultural climate.
>
> The signs are already here – kids are moving out to the farms and communes... A whole new culture is with us, together with a new zestful interest in customs and traditions of olden days. This is what 'The Country Bizarre' is all about.

It was the first underground magazine to receive a favourable mention from the newsletters of the two leading conservation groups, the Conservation Society *Newsletter*, and the Soil Association newsletter *Span*. This praise showed the common ground between the romantic views of eco-freaks and the traditional concerns of conservationists. Both centred on a nostalgic interpretation of the countryside as the spiritual and ecological home for British people.

The ConSoc *Newsletter* (July/August 1971) wrote that *Country Bizarre* had 'delightfully unsophisticated approaches to grimy, sordid issues and very pleasant information on country festivals and fairs, herb lore, records of birds sounds and gentle living. This magazine is the result of a very sane reaction to an aspect of Britain as it is now (and may not be much longer)'. *Span* (June 1971) commented that 'among the many new magazines more or less connected with the "under-ground" movement, the *Country Bizarre* is at once one of the most original and one of the most attractive'. Perhaps what appealed to *Span* (whose editor was Michael Allaby, by then a co-editor of *The Ecologist*) and what was remarked upon, was a contribution in the third issue by Edward Goldsmith entitled 'Stable Society' (as well as an ad for *The Ecologist*).

Country Bizarre was also the link between the old 1960s underground press and the new 1970s eco-press. It received glowing praise and support from *IT*, the only alternative paper to show interest in the concerns of *Country Bizarre*. This coverage was most certainly due to the influence of Joy Farren (a member of the *IT* editorial team) and copious writer for *IT* in 1971 on British mythology and ecology.[46] In *IT* no. 114 Joy Farren wrote: 'Unbelievably even better than ever. Please if you care for beautiful things, for the country, for people to be happy, buy *Country Bizarre*'. Starting from 1971 *IT* started running articles on ecology, wholefoods, herbs, and crafts, and undoubtedly it was influenced by the content and style of *Country Bizarre*. Thus it was able, due to its much larger circulation, to spread the ecological themes of *Country Bizarre* to a wide and mostly urban audience. Undercurrents, the alternative technology magazine, started in 1972 as *Country Bizarre* was closing also gave it rare praise describing it (in issue no. 3) as 'surely the best produced of all alternative mags'.

The Snail

One of the most successful and longest lasting of the eco-freak (and community) papers was *The Snail*, initially called the *North Devon Snail* for the first fourteen issues. Peter Blake (also a writer for *Country Bizarre*) started it in March 1971 as a local paper, but after two years he felt restricted by its 'localness' that limited what it could say.[47] Clem Gorman, a chronicler of the underground scene with his books on communes, reported that Blake 'wanted a paper that had the editorial freedom of a national underground paper like the London-based ones but carried responsibility for the underground Left culture of one part of England'. *The Snail* was a great success; within a year it had more than five thousand readers, and as Spiers commented: 'It grew from the paper of a locality to reflect the unsensational life of a region; its honesty, sincerity, and way of relating to people have given it a national circulation'. *The Snail* was concerned with the issues of decentralization and the revitalizing of the countryside, and sought to build practical alternatives through self-sufficiency and craft skills.

Anarchism and pacificism

Two underground magazines, *Dwarf News* and *Global Tapestry*, and one radical magazine, *Peace News*, gave space to ecological issues from the anarchist and pacifist perspective. *Dwarf News* was published in London, started in June 1971 and lasted until June 1973 producing about fifteen issues.[48] It was the newsletter of the Dwarfs, a group which started in early 1971 as the Aquarian Liberation Front with the aim of establishing an alternative society without government, authority or bureaucracy. It advertised itself in *Resurgence* as a movement for 'mobilizing the underground and creating a human-scale society of co-operation and love'. Very little has been written about the Dwarfs, but they were undoubtedly influenced by the Diggers, a well-known San Francisco group, and the Provos, the Dutch anarchist movement.

Global Tapestry was run by the committed pacifist and vegan, Dave Cunliffe,[49] who described the publication as 're-orientated towards alternative models of communal relationship, illuminated lifestyle and universal liberation. Interests include anarchism, ecology, pacificism and poetry'. Cunliffe produced his magazine from Blackburn, Lancashire with the first issue out in June 1969, then sporadically until 1972, and eventually it became more a series of occasional pamphlets than a magazine. Cunliffe had started contributing to *Resurgence* in 1966 and he was able to recruit other writers from it, including Roger Franklin who contributed a long two-part article on 'Ecological Action' in 1969-70.

Peace News, the long-established pacifist paper, was edited in the early 1960s by Theodore Roszak, later to find fame with his book *The Making of a Counter Culture*. It was not part of the underground press but according to Nigel Fountain it 'stood at the crossroads of mid-1960s dissident culture' and trailed the bulk of the underground's preoccupations for the next half decade. This included a concern with the new politics of ecology, communal living and vegetarianism. Its pacifist readership overlapped considerably with the underground, and thus it acted as a bridge between the old and new left groups. There was considerable overlap of writers; Dave Cunliffe for instance wrote for *Peace News* (25 August 1972) on the links between ecology and mysticism.

Communes

The communes movement was much smaller in the UK than in the USA, where it is estimated thousands of communes were started in the 1960s. This could be because of the relatively cheaper price of land and property and the laxity of planning restrictions in rural America compared with rural Britain. The self-styled 'Communes Movement' in Britain had *Communes* as its journal. The movement evolved from vegan and vegetarian communities in the 1960s. However the appeal of *Communes* was limited, with most left activists dismissing it as escapist and utopian. As Roger Lewis, the underground historian, commented in his book *Outlaws of America* (1972): 'The majority of people prepared to discuss communes seem to want retreatist enclaves where they can relax, eat wholesome food and forget the world outside. In consequence the revolutionary left has dismissed the drive towards collective living as utopian, and many of the less competent idealists who have embarked upon building communes have become rapidly disillusioned'.

Richard Neville was more sympathetic, saying that the agrarian commune movement can be seen as a retreat to an era when work was directly related to basic human needs even if 'their satisfaction is merely a variation on that sentimental respect all city dwellers have for those who till God's soil'. Michael Allaby in his book *The Eco-activists* devoted a whole chapter to 'The Communes', with long extracts from the *Communes* journal (on the Selene Commune) and also from *Hapt*, a free magazine on communal life and theory produced by the Hapt Diggers. As Allaby comments very sweepingly: 'The commune movement is clearly a part of the "underground", this amorphous second culture that is being created beneath the surface of and perhaps amid the ruins of western civilization... The commune movement itself is growing in this country as more and more young people reject the mainstream of society and seek to find for themselves an alternative life style'.

The rural commune ideal or perhaps mythology has had a very powerful influence in ecological and anarchistic circles, although the number of people living in communes in Britain was always very small.[50] The commune and the ecological movement shared an anarchist heritage, a belief in individual values and a tendency to spiritualism, besides caring about pollution and self-sufficiency. The rural ideal was the organic farming community, and some members of the Soil Association thus supported and encouraged the commune movement in the early 1970s despite the immense gulf in social attitudes and lifestyles. Michael Allaby contributed to Clem Gorman's first book *Making Communes* (1972), writing on organic farming and gardening. Other contributors included Sid Rawles, creator of the Digger movement in England in 1967, and founder of the Dorinish Island commune in Ireland.[51] The backgrounds of the two men could not be more different: Allaby was an ex-police cadet and fighter pilot turned journalist, while Sid Rawles was the leader of the Hyde Park Diggers who became famous for his participation in the occupation of 144 Piccadilly in early 1969. What united these two, and many others from very diverse backgrounds, was the desire to live off the land, and achieve a more self-sufficient and simpler way of life. The ecological vision thus appealed to both the 'freaks' and the 'straights', both seeking to escape the 'horrors of the industrial complex'.

The direct influence of the rural underground press and the 1960s communes was ephemeral, few of the communes still survive, and none of the little magazines lasted more than a few years. However their importance is as a bridge between the old organic tradition – as typified by the Soil Association – and the new environmental press, such as *Undercurrents*, that was to follow in the early 1970s.

The environmental press

The first of the new wave of environmental magazines was *Your Environment*, which ran from December 1969 to June 1973. It was started by Ted Hughes and a couple of other poets but it was later sold to a commercial publisher who lost interest in it. It was not part of the alternative press, as it was a well-produced conventional magazine, similar in style and content to the US magazine *Environment*. According to Walt Patterson who joined the magazine in 1970, becoming its co-editor in 1971 and eventually its editor, it had about two thousand subscribers of 'very high quality all over the world'. It carried articles by the leading environmental writers of the day, including Jeremy Bugler, Ian Breach and Oliver Thorold, all of whom were to be involved in the Windscale Inquiry in 1977. The first two, as journalists, provided sympathetic coverage of the opponents whilst Thorold was junior counsel for FoE. The magazine's influence is hard to assess, but undoubtedly important. It has been little studied by academics as only four libraries in the UK stock it; if it were not for Patterson's reminiscences of his involvement it would be largely forgotten.

The next magazine to be founded was *The Ecologist*, which started in June 1970. This was to become the best-known environmental magazine in Britain when it launched its *Blueprint for Survival* in January 1972 that attracted intense media publicity. Initially it was not considered part of the alternative press as it was in format a conventional magazine. It was however publicized in alternative magazines and by 1976 listed in the *Alternative Press Index*. Its *Blueprint* attracted little interest in the eco-press, being reviewed only in *Resurgence* (November 1972) and *Span* (nos. 57 and 58), with a critical letter from John Fremlin, a nuclear supporter, to *The Ecologist* (July 1972) disputing the *Blueprint*'s views on the dangers of radioactivity. The lack of direct interest in the *Blueprint* maybe reflected saturation with the press coverage, and jadedness with the word and the concepts of 'survival'.[52]

Undercurrents

The first of the eco-freak magazines was *Undercurrents*, whose first issue came out in January 1972. Its format for the first four issues was very alternative, a load of pamphlets and A4 sheets contained in a plastic bag, which emphasized its accessibility to any groups who wanted to contribute material. Like the *Whole Earth Catalog* it reflected a diverse (and perhaps contradictory) set of interests ranging from technophobic rural nostalgia and self-sufficiency to technophile love of electronic gadgets. I described it, in my *Alternative Technology Directory* (1978), as the 'magazine of alternative technology and science with emphasis on self-sufficiency, home electronics, free radio, workers co-ops. Good news coverage on energy, communications, industrial co-ops and the best in radical intellectual thought'.

Undercurrents was initially based in London, and its weekly editorial meetings attracted most of the movement intellectuals involved in the anti-nuclear power campaign. It was the first magazine to have a special feature of sixteen pages (in no. 9 in early 1975) devoted to nuclear power with the lead article entitled 'Why the NUCLEAR POWER PROGRAMME MUST be STOPPED'. From then on it gave extensive coverage to the anti-nuclear power campaign. *Undercurrents* had by far the largest circulation of any of the new environmental magazines. That, together with its longevity (it ran until no. 51 in March 1982), its network of influential London activists, and the publication of its book *Radical Technology* (1976) ensured its intellectual pre-eminence amongst the eco-freak press and the AT movement. Many of its contributors were later to enjoy careers as journalists, writers and academics, particularly in teaching renewable energy at the Open University.

Aether

Aether ran from 1973 to 1976 and was founded by students from the Environmental Protection Society at Aberdeen University. After three issues (and in October 1973) the editor Dick Firmin – together with his wife Rona and a few friends – moved to a croft near Aberdeen, in Scotland.[53] It describes itself as 'a bimonthly magazine produced at a crofting commune in Aberdeenshire, talking about ecology, survival and alternatives'. It also mixed local news with features on national and international groups and events. *Aether* continued in the tradition of *Country Bizarre* in being well-illustrated with original artwork. It had articles on the delights of crofting and rural life, but unlike *Country Bizarre* it developed a much sharper political interest based on anarchism, feminism and communes. As such it preferred the anarchist views of Murray Bookchin to the conservative views of Edward Goldsmith, and the People Party. In an editorial (July 1974), it defended communes from criticism by saying that it was 'important to set up bases in the country where individuals and groups may develop self-sufficiency and explore new ways of living and working together'.

Aether ceased publication in mid-1976, and thus played no part in the anti-nuclear campaign, though in its last issue (April 1976) it had an advertisement for the first 'week-end camp-in' at Torness that was one of similar protests organized by FoE groups at Windscale and Sizewell. The importance of *Aether* lies in its demonstration to its contemporaries that a small group of (ex-)students could launch a well-produced magazine with a national circulation, and in its promotion of a vision of self-sufficiency based on communal anarchism rather than Goldsmith's autocracy.

Seed

Seed – subtitled 'The Journal of Organic Living' – was produced from Ceres wholefood shop in London. It started in 1972 and ran until 1977, and played a major part in promoting interest in organic agriculture, wholefoods, and health and nutrition issues. In this it can be considered as a successor to the pre-war alternative health magazines, such as *Health and Life* which was edited by Edgar Saxon who also established Vitamin Cafes – the first health food restaurants – and a health centre in Wigmore Street, London, where he had a practice in 'natural health' medicine. However in contrast to other serious publications, *Seed* aimed for a lighter, more humorous approach. In its first editorial it said it believed that there was 'too much gloom and dark prophesy in the movement for natural living. We believe there is a brighter side and that soon, the masses of humanity will come to realize that natural living is not incompatible with joyful living. Our aim is to provide a journal which will blend the seriousness of our mutual efforts with the humour and satire that is an inherent part of it'.

Wholefoods, and in particular macrobiotic food, were considered with deep scepticism and as symbols of apathetic hippies by most writers, hence the derogatory use of the term 'brown bread and sandals'. Thus the AT book *Radical Technology* could dismiss *Seed* as 'the monthly organ of the macrobiotic and health food freaks' whilst giving the traditional organic groups, the Soil Association and the Henry Doubleday Research Association, much more respect. The format of *Seed* proved successful, and it soon established a much higher and wider circulation (5,000 by mid-1973) than the newsletters of these older groups and was thereby able to promote the message 'you are what you eat'. It thus acted as a bridge between the older organic movement and the eco-freaks, and established the old issues of diet, nutrition and organic agriculture as central to the new lifestyle environmentalism.

Whole Earth

This magazine was started in 1974 by a group of students at Sussex University, who had formed an environmental group in late 1973 called the Sussex Whole Earth Group – the title of the group was inspired by the *Whole Earth Catalog*. It was produced in Brighton from 1974 to 1979 and took its inspiration from *Country Bizarre* and *Aether* and like them placed great emphasis on original artwork. Initially it had articles on organic gardening, crafts, wholefoods and self-sufficiency, but later broadened out to include issues such as worker co-ops and wholefood shops, 'land for the people', energy issues, AT and free festivals. The first article on nuclear power did not appeared until no. 9 (early 1977); this was by Geoff Beuret from Hastings FoE group. This slow response to the then current debate probably reflected the dominance of *Undercurrents* and *Peace News* in reporting this issue, and the lack of any anti-nuclear protests in the south of England.

Like many similar, but shorter lived, eco-freak magazines it was part of a national network of groups based on communal living, co-operative working, radical politics, wholefoods and free festivals (for an encyclopedic listing and description of such groups see *Alternative England and Wales* (1975) by Nicholas Saunders). It was through such small magazines that the anti-nuclear message given out by national magazines, such as *Undercurrents*, *Peace News* and *The Ecologist* was disseminated to tens of thousands of environmentalists across the country.

Chapter 12

The environmental impulse

A s Walt Patterson so aptly put it in his first account of the British anti-nuclear movement in *Journal of the Institute of Nuclear Engineering* in 1979, 'What is an "environmentalist", and what is it about nuclear power which arouses "environmental" attention and concern?' An 'environmentalist' is commonly understood to be someone concerned about the 'environment' but that term is open to much dispute. The terms 'environmentalist' and 'ecologist', meaning someone concerned about and working for the protection of their environment, only came into usage about 1970, reflecting the use of these words by government and in magazine titles. People initially called themselves 'eco-activists' (as in the title of Allaby's 1971 book) but by the early 1970s most popular paperbacks used the term 'environment' or 'environmental' in their titles, and hence the usage 'environmentalists' or 'environment movement'. By the 1980s the term environmentalist had swept aside all previous rival terms, such as ecologist or conservationist, and established its meaning according to *The Concise Oxford Dictionary of Ecology* (1994) as: 'One who holds that damage to the natural environment resulting from human activity is so severe as to present a challenge to the survival of many habitats and ultimately perhaps to the continuance of life on Earth, and can be addressed only by major reforms of the way people live and industries function'.

The environmental impulse

Why do only a minority of people who say (according to opinion polls) they care about the environment, become activists? Why do most remain passive? What is it that galvanizes certain people to do something? These questions are the staple of most academics who seek to explain the rise of the environmental movement, and the explanations vary according to the discipline of the writer.

Political analysts talk about it in terms of 'opportunity structures' in political discourse and sociologists ascribe it to the rise of 'new social movements' due to changes in society. One widely subscribed view is that a change in values away from 'materialist' towards 'post-materialist' in the 1960s explained people's interest in ecological issues and was the cause of anti-nuclear attitudes. Others ascribe it to the reality of worsening post-war environmental conditions, particularly air and water pollution caused by new technologies.

The argument for a change in values as a cause of environmental activism rests on the idea that as people get richer they can afford to do something about a polluted environment. The rich, it is claimed, have different desires and aspirations from the poor who, it is argued, are willing to put up with (and generally ignore) pollution and a low quality environment for the sake of jobs and prospects of material possessions. This view is however being challenged by the 'environmental justice' movement.[54] The values view was endorsed by Samuel Hays, the veteran US environmental historian in his book *A History of Environmental Politics*, when he wrote: 'At root, environmental affairs in modern life are a combination of changes in values and environmental circumstance'. He argued that these new values came from increasing post-war affluence and the willingness to pay for a clean environment. However the cause of these new values is experiencing concrete examples of environmental degradation in everyday life, such as pollution, the destruction of natural resources, and overcrowding.

There can thus be seen an element of material (or vested) self-interest in this campaign to 'clean up the environment', which attracted criticism from the left during the 1960s and 1970s. This concern for environmental issues as class-based was famously articulated by Anthony Crosland in a speech in 1970 – reprinted in a Fabian society pamphlet *A Social Democratic Britain* (1971). Speaking of sections of the conservation lobby, he said: 'Their approach is hostile to growth and indifferent to the needs of ordinary people. It has a manifest class bias and reflects a set of middle- and upper-class value judgments [for which] preservation of the status quo is the sole consideration'.

That most members of environmental groups (in the 1970s) were middle-class is incontestable. Sociologists in the 1970s saw environmental activism and opposition to nuclear power as the product of the emergence of a 'new class' based on middle-class professionals working in the service sectors who were, through environmental protest, attempting to gain improved social status and greater economic power. Undoubtedly many of those involved in anti-nuclear protest were members of the 'new class' but better-educated people find it easier to become active than do other elements of a population. Also 'new class' professionals because of their media skills can become influential people, and such people can play a leadership role in a wide range of social movements – new, radical, reactionary or otherwise.

Attitudes to nature

The key driver to environmental activism may be attitudes towards the protection of nature. This dates back to the Victorian concern for wildlife protection, which stemmed from two powerful forces – the strong enthusiasm for natural history and the crusade against cruelty to animals. It was then that the first animal protection and nature preservation groups emerged – the Society for the Protection of Animals in 1824 and the Commons Preservation Society in 1865. Underlying the founding of all the groups devoted to the protection of nature was a fundamental dispute about the future direction of society that rested on attitudes to nature versus culture (see Figure 1). In Western thought this grew into polarities such as countryside versus city or wilderness versus civilization and these implied a contrast between wild and controlled things. In personal terms nature was often associated with intimacy rather than formality, with instinctive, 'natural' impulses rather than self-control and planning and, in brief, with feelings as opposed to logic.

Below shows a diagram of ideas, common in modern thinking, distinguishing nature from culture.

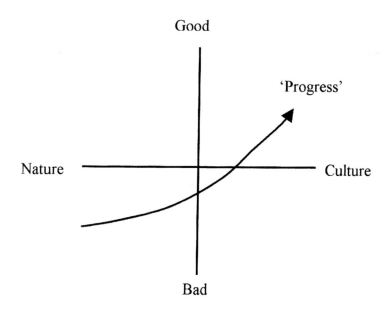

According to Weart either extreme could be seen as good, bad or a mixture. In the lower left of the diagram we would find the old view of wilderness as a thorny forest full of wolves and demons; in the upper left, an unspoiled fruitful Arcadia; in the upper right, an orderly utopian White City; and in the lower right, a robotic slave state. The traditional arrow of 'progress' or economic growth is shown as an ideology of replacing lawless wilderness with beneficent civilization. Opposition to this view of 'progress' was at the core of antinuclear ideology.

These patterns of association have become so pervasive in Western thought that, as Spencer Weart remarked in *Nuclear Fear*, most people take it for granted:

> Nature was to culture as wilderness was to civilization, wild to self-controlled, victim to authority, feelings to logic, and female to male, not to mention liberty to order, freedom to security, Dionysian to Apollonian, organic to mechanical, 'soft' to 'hard' and so forth indefinitely.

These associations carried a scale of values. Nature and culture were generally seen as equally capable of good or bad, and the ideal state was envisaged as the harmony of human activity (or civilization) with the natural environment. However another way of thinking, that of mastery and control of nature, has also been long-dominant with emphasis on man's duty to transform the useless disorder of desert and forest into an organized landscape. By the nineteenth century this idea of 'progress' was universally identified with the growth of science, industry, and social order, but was increasingly criticized by intellectuals who began to praise untouched wilderness along with spontaneity and freedom.

This association of nature with feelings and emotions, rather than intellect, is implicit in the ideas of Peter Hay, a green philosopher who in his book *A Companion to Environmental Thought* believed that environmental participation, or what he called a 'green commitment', is not derived from a theoretical or even rational perspective. He argued that this 'green commitment' – or what he calls the 'ecological impulse' – is due to an instinctive ecological compassion often caused by 'some trigger or impulse'. He wrote that 'the wellsprings of a green commitment – at both the activist and more passive levels of identification – are not, in the first instance, theoretical; nor even intellectual. They are, rather, pre-rational'. Hay described this 'pre-rational impulse' as a deep-felt consternation at the scale of natural destruction wrought upon the 'increasingly embattled life-forms with which we share the planet'. Thus to Hay 'the act of instinctive ecological compassion' at 'witnessing the obliteration of once-abundant life' seems to be the most potent source of green recruitment.

This observation of the importance of the massacre of wildlife to activism is borne out by the historical record on the founding of conservation groups specifically devoted to the preservation of fauna and flora. One of the earliest campaigns was against the shooting of seabirds at Flamborough Head, which led to the passing of the Sea Birds Preservation Act of 1869. The environmental movement has always been campaign led, a popular response to an assault upon people's lives through pollution or the destruction of nature, and the early campaigns reflected this. This movement had of course many strands, from conservative to revolutionary, and as a journalist remarked in *New Statesman* (3 August 1973:

> The main strength of the movement is in fact in its diversity. It is the people who react spontaneously to protest some specific exploitation of their own areas or disruption of their lives who are forcing the political establishment to take environmentalism seriously, and continuing to challenge ideas which until recently were taken for granted: the dominance of man over nature, and the inexorable progress of technology.

Impact of suburbanization in the United States

Samuel Hays, in *A History of Environmental Politics*, argued that the environmental movement is at heart a consumer movement keen to secure and protect a more satisfying environment for urban dwellers. He believed an important factor was people experiencing areas of outstanding natural beauty such as in National Parks. These became more accessible due to the post-war growth in leisure opportunities and motoring holidays. Hays believed that more people become engaged with activism through this tourism than by reading Henry Thoreau or John Muir – although these influences may come later as people search for wider meaning and explanation. Hays' views are echoed by Adam Rome in his study of the impact of post-war suburban sprawl in the book *The Bulldozer in the Countryside*. Rome argued that the desire to preserve wilderness area, such as Echo Park in the early 1950s, 'was the tip of an iceberg' and that far more influential on people was the environmental consequences of suburban development. He wrote: 'Again and again, the destruction of nearby open spaces robbed children of beloved places to play – and the losses hit home more vitally than the threats to far-off sites like Echo Park ever could'.

Another author, Langdon Winner, in the final chapter of his book *The Whale and the Reactor* (1986), traced his moral outrage over nuclear technology to the degradation of the Californian countryside dating back to childhood days. He wrote:

I remember that at a fairly young age I began wondering why it was that so many of the things presented as improvements were actually inferior to the things they replaced… Attractive wooden buildings that showed the character of human imagination and the carpenter's hand were replaced by flimsy stucco tract homes and apartment buildings. Land that was among the most fertile in the United States was paved over for highways, shopping centres, and motels. Even the roads, which in their primitive two-lane versions had paid some respect to the contours of the hills and placement of older houses and farms, with the coming of high-speed four-lane freeways, began to slice directly through whatever stood in their path: groves of ancient oak trees, whole neighbourhoods, landmarks a hundred years old. "That's progress," people would say; or "You can't stop progress."

This account of the impact of the destruction of the Californian landscape by suburbanization is mirrored in the science fiction novel *Antarctica* (1997) by Kim Stanley Robinson. Here the green activist hero, Wade, is described:

He grew up near Redlands, California in a region still covered by orange and lemon groves, and avocado orchards. But the inexorable metastasizing of Los Angeles rolled over his home when he was a boy, and he watched mute and uncomprehending as over the years of his childhood and youth, the groves were cut down and replaced by freeways, malls, condominium complexes and gated suburban communities. And when he went off to Berkeley to go to college, and began to think about what had happened to his childhood home, it made him mad.

Wade gets a law degree and joins the Wilderness Defense Club, but sickened by their compromising, quits and joins an informal extremist sabotage group called 'ecotage international' (an underground version of Earth First!).

Nature destruction in Britain

The British conservationist William Adams similarly argued in his book *Future Nature* that the sense of lost nature, and feelings of helplessness in the face of vast forces for change, are powerful stimuli for conservation. He believed that concern at change in particular places or the destruction of particular individual creatures can prove radicalizing and gave an example from his childhood:

I can certainly recall one particular event in my own childhood that stands out for its effect on my attitudes to nature, and to the ways people used and abused it. It happened when I was quite young and new houses were built in an empty plot opposite my house. Doubtless I idealise the scene, but there was a row of red horse chestnut trees standing above the road behind an old wall, and in my memory they are vast and seemingly immovable. In spring they were resplendent in bright leaves and bold candles of blossom, and at night tawny owls called from their branches. When the old garden in which they stood was sold for housing, the trees were felled. By a terrible coincidence, the work was done in May, when the trees were in full flower. Afterwards, when the chainsaw had stopped and the workmen had gone, there was nothing remaining of the trees at all, except the roadway strewn with red petals.

Adams can still recall his sense of outrage and pain at this event: 'To do this to trees (my trees) was bad enough, but to do it in spring seemed to me a criminal act, an offence to nature and to human decency. To me this was a symbol, a worked example, of humanity at its most outrageous: greedy, selfish, thoughtless and uncaring. Unreasonable as my feelings might have been, they reflected my shock at the impact of normal day-to-day life on nature'. The conflict between what seemed the self-evident values of the trees and the casual indifference of those who planned the houses set a chain of thought going in his mind, and such experiences turned him from being simply someone who like birds and bugs into a conservationist.

Similar experiences were found by Derek Wall in his interviews with Earth First! activists, many of them saying that they had been influenced into green activism by their childhood experiences of the destruction of the countryside. As Wall observed, 'transgression of a place cherished in a childhood memory' may later fuel activism while Adams remarked: 'People who care about nature can often identify similar moments when their feelings crystallised, and they began to understand the effects of industrialised society on wildlife and the countryside'. Thus action, in the form of the environmental movement, often came from witnessing the destruction of nature and the appreciation of scenic wilderness, not conversion through the reading of conservation texts, though it may be subsequently justified by recourse to an intellectually generated system of ideas.

Social movements and intellectuals

Little can be done about environmental destruction by the individual acting alone. Most people realize that to be effective they have to band together to form campaigning groups, which can then combine to form a mass movement. As sociologist Alan Mazur remarked in his book *The Dynamics of Technical Controversy*: 'The options usually available to someone who wants to express his concerns are limited: either waste one's efforts on solitary protest with little chance of success, or join a currently-running protest movement and pool resources with other sympathetic souls'.

There is a great deal of research and academic debate on the significance of 'social movements' of which the anti-nuclear power and environmental movements are considered examples. They can be defined as loose networks of individuals and groups who have no organizational affiliation, and who are engaged in collective action motivated by a shared concern about environmental issues. The emphasis is on networking, participation and collective action by a wide range of organizations, groups and individuals.

A social movement has four typical characteristics:

- It is based upon informal networks. These may include more formal organizations, such as pressure groups, but are also broader than them.
- Those involved must share a set of beliefs and collective identity. This defines whom or what they see as allies and opponents, what their goals are and how they are to be reached.
- Social movements are involved in collective challenges and may threaten their opponents with sanctions.
- Social movements use protest and cultural practices, which may or may not be confrontational.

Research into social movements by political sociologists focuses on the cultural and sociological dimensions of collective action, and tries to relate this to structural changes in society. In contrast, political scientists describe social movements as interest or pressure groups, and focus on the groups' involvement in policy-making. Most early literature on the anti-nuclear campaign has taken the latter approach, concentrating on those groups and individuals whose contribution had an impact on the UK nuclear policy debate, while more recent literature assesses the success of the international anti-nuclear movement in terms of policy influence. Such a perspective gives emphasis to those groups (such as FoE) and individuals (such as Walt Patterson) with the required expertise and 'respectability', that can take part in national policy debate conducted in the national media and in parliamentary and judicial inquiries.

In contrast, a political sociology approach concentrates on the actions of local anti-nuclear groups in developing a movement culture and mobilizing support and places emphasis on the creation of activist identity and new forms of movement organization and expression. These include the rejection of 'hierarchical, authoritarian models of social organization' in favour of 'decentralized consensus decision-making, non-sexist practices and non-violence'. In wanting radical changes in society social movements are also unafraid of asking utopian questions. As Ron Eyerman and Andrew Jamison comment in their book *Social Movements* (1990), social movements as bearers of new ideas bring societies back to the big questions of 'what is man? what is nature? what is history?'

Movement intellectuals

I am concerned with the interplay of ideas and action, with how some ideas can create a social movement. Crucial to this creation is the role of some people, termed 'movement intellectuals', who can not only communicate previously obscure ideas to a mass audience, but also build a movement – a mass of supporters willing to act – where none existed before. In sociology intellectuals are viewed as those who are professionally engaged in the production of ideas or the manipulation of symbols – or in new class terms as 'producers of culture'. To environmental historians such as Samuel Hays, they are those people who write, read, absorb and are preoccupied with ideas, and who are 'part of a network of writers, publishers and book reviewers who sustained mutual ways of thinking'.

Eyerman and Jamison believed a necessary condition for the formation of a social movement is the generation of movement intellectuals who 'articulate the collective identity that is fundamental to the making of a social movement'. They distinguish between the role of movement and establishment intellectuals in the formation of social movements, with the latter crucial in laying the groundwork for the formation of the movement. As Eyerman and Jamison commented:

> The environmental movement emerged in large measure from the activity of estab-
> lished intellectuals, thus seemingly giving support to traditional accounts of the role
> of intellectuals in social movements. It was out of the writings of ecologists and
> conservationists and perhaps especially popular science writers that the cognitive
> identity of environmental activism first came to be articulated. They were, for the
> most part, persons who were already socially legitimated intellectuals, either
> through their academic positions (Barry Commoner, Paul Ehrlich, René Dubos,
> George Borgstrom) or through their popular writings (Rachel Carson, Lewis
> Mumford, Vance Packard).

Establishment intellectuals however did not create the environmental move-
ment, for this was done, in its early phases, by movement intellectuals through more or less traditional means of mobilization – by creating sense of collectivity through founding their own organizations and networks. The movement intel-
lectual articulated this new collective identity through speeches, tracts, articles and books. Central to this process is the creation of an 'Other' against which the budding movement will interact. As Eyerman and Jamison commented: 'The Other is not merely an intellectual construction, but is almost always a real social actor, an authority, the government, an institution, the state, or a conglomerate of individuals, the "technocrats" with whom the movement must strategically

interact'. With the anti-nuclear power campaign this 'Other' was the controlling nuclear institutions, in the UK the Atomic Energy Authority (AEA) and in the USA the Atomic Energy Commission (AEC), and more generally what Robert Jungk called the 'nuclear state' in his book of the same title (1979).

As the movement developed, the roles of movement intellectuals expanded from facilitator and ideologue to mass media communicator and finally to counter-expert. These counter-experts were often professionals based in existing institutions. They were critical of the decisions and standards of the government experts, and were willing to challenge them on behalf of what they saw as the 'public interest'. The counter-experts opposed not merely the state and its environmental bureaucracies, but also the elitist idea of knowledge represented by the state expert. They symbolized a democratic ideal and a sharing of knowledge. Such intellectuals, with the credentials and skills gathered in the institutions of the established society, played a crucial role at the public inquiry, where the ideologue was at a disadvantage from lack of expertise. Thus Walt Patterson, with his technical training in nuclear physics, fared well at the Windscale Inquiry whilst ideologues such as Edward Goldsmith and Charles Wakstein received adverse criticism.

As the movement matured it became increasingly professional and organizationally based, often employing the former students who had taken an active part in earlier campaigning and been the spokesmen. Later, after the movement had dissolved or been incorporated into institutions, they took influential positions in university departments, in law and journalism, and in professional organizations. They made history and then they (mostly men) wrote about it, and their accounts and analysis of their movement days form the basis of this (and many other) accounts.

Why an activist?

Why do individuals join groups? Joe Weston – an ex-FoE activist – in his report *The F.O.E Experience* found that a large percentage of FoE supporters were at college during the 1960s and were influenced by the events of that decade. These influences included rock music, television images of war, famine and environmental disasters, and certain books that helped steer them towards green politics – Rachel Carson's *Silent Spring*, Ehrlich's *The Population Bomb*, and Rene Dubos' *So Human an Animal*. Such books he believed may have first 'turned them on' to the problems of the environment. Of course not every 1960s student who got 'turned on' to environmental issues joined FoE or other groups. Many who joined a group remained passive members, only paying a subscription and taking little part in activities, and only a minority became activists. Motivation for activism was very varied. Weston reported that his Oxfordshire FoE group was a mixture of individuals who came together for a diverse set of reasons but were able to work together on specific events. He found that there was never any discussion of the reasons for the campaigns or the events, neither was there an attempt to draw up a constitution or explanation of why the group believed in what they were doing. He stated: 'They all believed that it was wrong to continue to kill whales, to dump nuclear waste and to destroy wildlife sites. But simply because they shared those beliefs does not mean that they necessarily shared anything else'. However it was the belief in (or enjoyment of) activism, not environmentalism, that brought them together: 'Some were natural campaigners who could find and justify a campaign for almost anything. Others just wanted to be part of the general "buzz" that surrounded such activity and of course still others did believe in green or ecological politics'.

In his interviews with Earth First! activists Derek Wall found many different reasons for participating: 'Accident, friendship, personal conviction, pleasure and political calculation were just some of the factors cited as influential'. However the two common features, he found, were either a long-standing green commitment, often of a passive nature, or activism outside the green movement. Wall argued that people become activists by being recruited by their friends to participate in activities. He believed that activism is a way of life, and that new recruits are introduced to it by their friends through gradual involvement in 'direct action', moving over time from the fringes to the centre. As he commented: 'Few individuals jump onto bulldozers... when first involved: repertoires have to be learnt. Activism, even in its most serious form, is a method of performance that must be developed and improvised'.

Over time as people become more experienced they may find direct action exciting or even enjoyable, and this, together with new friendship links, intensifies the peer pressure to become full-time activists. Wall pointed out that full-time activism depends on personal availability, and can only be for the small minority of politically concerned who lack work and family commitments. These people are likely to be young and free, either unemployed or students or more rarely self-employed. Participation in groups is however essentially a social process with many cultural overtones; the availability of free time and a grudge is not sufficient for successful involvement. Wall observed that 'the protestor' requires a cycle of gradual involvement and strengthening of network ties – 'direct action may be like a ritual that eases the often traumatic and anxiety-inducing passage from one identity to another'.

Whilst being a direct action activist can be a full time occupation, involving a change in lifestyle, most campaign groups – like FoE or ConSoc – cater to those with a conventional career seeking an ethical and exciting hobby. As Weston remarked on participating members of his Oxfordshire FoE group:

> For the most part though their involvement… was like any other spare time activity taken on by people everywhere. It was a sort of ethical hobby taken up by people who want to belong to a group, and who, at the same time, like to confront authority. Being part of an active, successful, campaigning, as opposed to life-style, FOE local group is almost a middle class equivalent of football hooliganism.

Women activists

Weston's account of environmental activism as a sort of 'ethical hobby' or a 'middle-class version of football hooliganism' perhaps reflects the dominance by men of the leadership and written accounts of environmental groups. This neglect of women activists in the UK reflects the 'sexism' of British protest movements in the 1970s. Women did the mundane work – the newsletter typing, the leaflet distribution, the petition gathering, the organizing of people to attend demos – while men did the more glamorous and high profile tasks – the writing of articles, speaking at conferences and giving interviews. There was a view that men were the activists, women the supporters. But beyond this simple 'sexism' there was in the anti-nuclear campaign an implicit assumption that only those with technical knowledge – preferably a Ph.D. in nuclear physics – could speak competently and credibly about scientific issues. There were some exceptions, notably Alice Stewart and Rosalie Bertell, academics both working on the health effects of radiation. But then health aspects were perhaps viewed patronizingly by men as acceptable concerns for women. When it came to the strategic, technical and policy issues, in general it was held that only those with expertise and rationality should take part in public debates. Otherwise there was the risk of being labelled 'emotional' and 'lunatic' and your views discredited, an opinion explicitly expressed by Walt Patterson in his interviews and books.

This view that men expressed 'facts' while women expressed 'emotions' permeated scientific thinking, even amongst the radical scientists who opposed nuclear power, such as those that wrote for *Undercurrents* – a magazine supposedly imbued with radical views on society. The views of women activists were ignored and their activities marginalized in favour of long 'scientific' articles by men. Only in autumn 1978 did *Undercurrents* give space to women writers, in its 'Special Issue on Women and Energy'. The only article on nuclear power was by Irene Coates, a member of the Windscale Appeal and participant at the inquiry. The introduction to her article (written by the women editors, presumably) fully expresses the prevailing views about women and emotions. 'We have heard the arguments for and against Windscale, most of them written by men. It is hard for a woman, faced with her feelings about nuclear energy, to be listened to seriously. A feeling response to the environment is as valid a starting point for protest and political change as an argument based on intellect or economic analysis. It is time that such responses are given the credibility they merit'.

This neglect of women's views was despite their forming a substantial part of the anti-nuclear power movement. According to opinion polls, women in all countries, in all time periods, were consistently more anti-nuclear than men. This phenomenon, Weart believed, was connected with a tendency, when any technology was mentioned, for women to think in terms of safety, the environment and their children, whereas men would think more in terms of the benefits and perils of scientific progress. Women diverged from men far more on nuclear reactors than on any other technology.

Given this anti-nuclear attitude it is not surprising that many of the activists in the early to mid-1970s against nuclear power in Britain were women – Betty Gazard at Stourport, and Jane Pink, Beryl Kemp and Irene Coates from ConSoc. This grass-roots leadership by women is also reflected in the United States, where roughly one-third of local opposition groups were led by women and women activists had a much higher public profile in campaigns. The activists Rose Gaffney, Jean Kortum and Hazel Mitchell played significant roles in the first Californian campaign against nuclear power at Bodega Bay in the early 1960s – see Thomas Wellock's book *Critical Masses* (1998). The group Another Mother for Peace, formed in 1967 by fifteen Beverly Hills women, mobilized young mothers in the early 1970s to campaign against the genetic hazards of radiation.[55] Then the 1970s anti-nuclear campaigns produced well-known women activists, including Anna Gyorgy who went on to write the seminal anti-nuclear movement guide book *No Nukes: Everyone's Guide to Nuclear Power* (1979), Sheryl Crown, active in the Clamshell Alliance who wrote *Hell No, We Won't Glow* (1979), and Helen Caldicott who wrote *Nuclear Madness: What You Can Do* (1980).[56]

Chapter 13

The environmental movement

The emergence of the environmental movement in the late sixties and early seventies was due to a combination of favourable factors along the lines of the cliché of 'the right person, in the right place at the right time'. The people were large in numbers and young and idealistic, particularly students, looking for a cause to champion. The places were the most liberal cities of the (Protestant) western world: San Francisco, New York, London, Berlin and Paris, where there were large universities. The time was 'the Sixties', an era of political turbulence, of protest and of conflict, and of divisions within ruling elites over moral and social issues. So both the opportunity and the population existed to create a social movement. But why did it turn to ecological issues?

Although there has been protest over environmental concerns for over a century there has seldom been a straight forward continuity of views, and it is historically inaccurate to imagine that Victorian activists – male or female – shared the same motivations and beliefs as 1970s environmentalists or 1980s greens. Though some may have been pacifist, anarchist or vegetarian, others held views which we would today denounce as highly sexist, racialist and elitist. Thus Peter Coates in his book *Nature* remarked that Derek Wall's attempt – in *Green History* (1994) – to identify a green lineage can only be done by excluding people with 'right-wing associations', such as Octavia Hill, the founder of the National Trust. As Peter Coates remarked: 'For many eco-socialist, the British National Trust is tarnished by its close ties with the aristocracy and its preoccupation with preserving and restoring the elite landscape of the country estate'.

It was however this influential minority of intellectuals and upper class Victorians who established the first groups devoted to nature conservation, the preservation of buildings, the protection of animals and landscape, and combating pollution. There was considerable mutual co-operation and support between the different societies, and overlap in membership and leadership. Wildlife and historic preservationists often employed the same terminology, such as frequent references to the relics of the past and the relics of nature, or to ancient monuments and natural monuments – and in much the same way the term 'conservation' became stretched to cover wildlife protection, building preservation and the husbanding of resources. The emphasis, as the titles of the organizations suggest, was on preservation and protection of old England against the corruption of 'progress' and the evils of the industrial city. In their book *Environmental Groups in Politics*, Philip Lowe and Jane Goyder remarked: 'Victorian preservationism was distinctly a gentlemanly avocation pursued by cultured people well removed from, and indeed averse to, the base pursuits of trade and manufacture'. It was also, in its provision of free open spaces in cities and the countryside, a response by social reformers, such as Octavia Hill, to head off the challenge by the industrial working class over access to the sporting estates of the landed aristocracy.[57] Overall in the first half of the twentieth century the emphasis was on 'conservation' right up to the founding of the Conservation Society in 1966 (see Table 2).

Group formation

Societies and groups are being started all the time; few however will last for decades or even a century as has the National Trust. The founding of groups is generally in response to public awareness and favourable media attention to an issue. Groups brave enough to campaign on an obscure (or unpopular) issue attract few supporters and little funding. Lowe and Goyder divide national groups into two categories, 'emphasis' and 'promotional' groups. The former are groups whose aims do not conflict in any clear-cut way with widely held social goals or values, while the latter promote causes involving social or political reform. Of the 77 groups they surveyed in the late 1970s, 48 were emphasis groups (such as the National Trust) while 29 were promotional (like the Conservation Society or FoE). They remarked that promotional groups were much younger (with a mean age of eight years) than the emphasis groups (mean age 43 years), with a tendency for groups to evolve from a promotional to an emphasis role, or to fade away if their reforming efforts seemed no longer relevant. Promotional groups were also much smaller, perhaps because they were so much younger, seldom having a membership of 10,000 compared to memberships of over 100,000 for long-established groups such as the National Trust or RSPB.

Most groups remain very small-scale and obscure, generally run by a committed enthusiast attracting few supporters and then fade away – the archetypal 'one man

Table 2: Date of founding of major 'environmental 'groups illustrating change in terminology.

Society for the Protection of Animals 1824

Commons Preservation Society 1865

Association for the Protection of Birds 1870

Society for the Protection of Ancient Buildings 1877

The Selbourne Society for the Protection of Birds, Plants and Pleasant Places
     1885

Society for the Protection of Birds 1889

National Trust for Places of Historic Interest or Natural Beauty 1894

Society for the Preservation of the Wild Fauna of the Empire 1903

British Empire Naturalist Association 1905, then renamed British Naturalist
     Association

Society for the Promotion of Nature Reserves 1912; then renamed Society
     for the Promotion of Nature Conservation

British Ecological Society 1913

Council for the Preservation of Rural England 1926

International Union for the Protection of Nature (IUPN) 1948, then
     renamed International Union for the Conservation of Nature and
     Natural Resources (IUCN) 1956

Conservation Corps 1959, in 1970 renamed The British Trust for
     Conservation Volunteers

World Wildlife Fund 1961

Natural Environment Research Council 1965

Conservation Society 1966

Committee for Environmental Co-ordination (CoEnCo) 1969

Friends of the Earth UK 1970

Greenpeace (London) 1970

Your Environment 1969

The Ecologist 1970

Socialist Environment Resources Association (SERA) 1973

Ecology Party 1975 (previously called People Party 1973)

Society for Environmental Improvement (SEI) 1974

Liberal Ecology Group; Conservative Ecology Group 1977

Greenpeace UK 1977

Green Alliance 1978

band'. Some are only designed for a short life span, such as local campaigns to oppose an unwelcome development, which include the classic 'NIMBY' protests. Finally there are a few societies which are initially successful attracting thousands of members and becoming nationally well-known, but then enter a period of long decline as the public loses interest in them and their members drift away to more dynamic groups. These societies are basically unable to adapt either organizationally or intellectually to changing circumstances and public interests. A prime example of this type of society was the Conservation Society (ConSoc) founded in 1966 to campaign on population issues; it peaked in 1973 and then entered a long period of decline before being wound up in 1987. Lowe and Goyder remarked that 'many groups which challenge dominant values never become established. Some simply cease to exist. This is a particular tendency with single-cause promotional groups. Sometimes it is demonstratively clear that they are redundant, having decisively succeeded or failed to achieve their original objects, and it is possible therefore for all concerned to agree to call it a day'.

One notable example of a failure was the Anti-Concorde Project, basically a 'one-man band' led by Richard Wiggs which achieved substantial publicity and support from environmental groups, including FoE. Whilst most groups packed up in a few years if they did not achieve some success, a few persevered for decades generally in obscurity often braving periods of media ridicule and hostility, with even sympathetic observers doubting their value. Lowe and Goyder commented: 'Only a few groups soldier on with little prospect of success. Examples include the Soil Association... and the animal rights and anti-vivisection groups'. In fact, these three groups have done particularly well since the early 1980s, with the Soil Association benefiting since the late 1980s from public fears over food safety, together with high profile support from Prince Charles and other conservationists. Animal rights campaigners have through dedicated, persistent and sometime violent direct action achieved major changes in farm welfare standards, often in the face of hostility from the media and prosecution from the police. These sorts of uncompromising activists are people who, in their life and work styles, are singularly committed to their values, which helps explain their tenacity in the face of adversity.

The new environmental movement

The media attention on environmental problems in the 1960s, on issues including the killing of African wildlife, the threats of over-population and the destruction of British towns led to the formation of such societies as World Wildlife Fund (latter called World Wide Fund for Nature or WWF) in 1961, the

Conservation Society in 1966, and hundreds of civic amenity groups. As the *New Statesman* (3 August 1973) commented, there was then a sudden burst of new enthusiasm for protection of the earth's ecosystems, what it called a 'fragmented crusade of maddening complexity but apparently limitless energy' which was simple in comparison with trying to describe what it stood for. 'What makes the environment movement so difficult to pin down is that its intellectual origins are as diverse as the motives of the people who have taken it up'.

Media attention and public interest also caused existing organizations who were not founded for environmental reasons – such as the Boy Scouts, the Church of England and the National Union of Students – to become interested and set up specific groups to cater for their members' interest. Environmental interest was also opportunistic, to show concern in order to recruit new members and, for business organizations, to create a favourable impression amongst the public – what is now labelled 'greenwash'. Throughout the 1960s an increasing range of organizations attended the three 'Countryside in 1970' conferences, from 90 for the first in 1963 rising to 335 for the last in 1970. However only a small proportion, perhaps 10%, of the increase in number of organizations attending the conferences from 1963 to 1970 was due to the formation of new environmental groups – in the decade of 1966 to 1975 fewer than 25 national groups were formed. Thus the vast majority of the increase in numbers was due to existing organizations and businesses claiming a new interest or concern about conservation. Similarly there was a big increase in the number of local amenity societies during the 1960s with 300 of them in 1960 rising to over 700 in 1969 and 1250 in 1975, with membership rising from less than 50,000 in 1960 to about 150,000 in 1969, and 300,000 in 1975. There were estimated (for 1979-80) to be about two million members of national groups with another one million in local groups. Only a small proportion of these millions were however to be active in the 1970s environmental movement, and an even smaller proportion in the anti-nuclear power campaign.

To the young eco-activists in the early 1970s there was a very clear distinction between the old conservation groups and the new groups. This new eco-action movement was very selective in the groups it wished to associate with, and only those older groups whose work reflected the new ecological values were included. This becomes apparent in a survey of the groups listed in the alternative and environmental press for the years 1968 to 1974. Those mentioned in the listings columns reflected the compilers' view of what constituted the environmental movement. Sometimes the juxtaposition of groups was strange – The Conservation Corps with the League Against Cruel Sports (*Country Bizarre* no. 3), or the Dwarfs and the Environmental Consortium (*Ink* no. 7).

There are 81 mentions of 33 groups in the listings columns, with by far the most frequent being the Dwarfs (13), followed by FoE (7), ConSoc (6), and Soil Association and Conservation Corps (5). Groups mentioned can be divided into 6 categories:

New groups (22): FoE (7), ConSoc (6), BSSRS (2), SERA (2), ITDG (1),
    Transport 2000 (1).
Eco-parties (21): Dwarfs (13), Diggers (3), People Party (3), Provos (1),
    Movement for Survival (1).
Traditional (16): Conservation Corps (5), Environmental Consortium (3),
    RSPB (2), CPRE (1), Nature Conservancy (1), National Trust (1), NSCA
    (1), TCPA (1), WWF (1).
Local (8): PEST (4), Greenpeace (London) (2), Action for Survival (1), Earth
    First (1).
Organic (7): Soil Association (5), Henry Doubleday Research Association (2).
Animal welfare (7): RSPCA (3), People's Dispensary for Sick Animals (2),
    BUAV (1), League Against Cruel Sports (1).

The large number of mentions of the Dwarfs reflected the tendency of the alternative press, which mainly mentioned them, to have listings columns and more frequent publication.[58] The popularity of the Conservation Corps (founded in 1959) probably reflected its appeal to young students, with many student eco-action groups acting under the name of the Conservation Corps.

It is interesting to compare this listing with that contained in *The Environmental Handbook* (Barr 1971) published by FoE. All the ones above in the 'traditional' and 'organic' categories are mentioned, including FoE and ConSoc, in the 'new groups' but none in the 'animal welfare'. There is also another listing of over 130 groups in the book *The New Battle of Britain* (1972), by H. F. Wallis, of which ten are new groups, six are organic, and six are animal welfare. There are also two groups concerned with birth control and abortion, reflecting conservationist preoccupation with population issues. The most respectable (or least hippie) environmental magazine *The Ecologist* listed just traditional groups plus BSSRS (in the December 1970 issue), whilst the alternative press always mentioned the Dwarfs and some of the 'new groups' and 'animal welfare' ones, but hardly ever the 'traditional' ones. It was *Country Bizarre* that acted as a bridge between the old and new groups, mentioning the Conservation Corps, RSPB and WWF alongside FoE and ConSoc, besides the RSPCA and the League Against Cruel Sports. *Resurgence* was also a bridge between the 'new groups' and the 'organic' agriculture groups of the Soil Association and the Henry Doubleday Research Association (HDRA).

Pre-FoE eco-action

In the early 1970s the venues for young eco-activists were somewhat limited. For the outdoor types there was the hard physical work of the Conservation Corps, for hippies the anarchist festivals of the Dwarfs, and for the serious (and dull) the local branch meeting of the Conservation Society. It was the good fortune of FoE that it was able to harness the energies and distinctive style of local groups that sprang up after its widely publicized bottle dump in May 1971. These groups, initially unwanted by FoE, combined the best points of previous eco-action groups – the hard work of the Conservation Corps, the street theatre of the Dwarfs and the intellectualism of the Conservation Society. However these local FoE groups consisted not just of new activists, they also drew on the expertise and experience of activists who had been involved in environmental campaigning far longer than FoE's full-time staff, who were ex-students. Also many were not content to follow FoE's centralized approach to campaigning; they wanted a more life-style led approach to bringing about social change.

The multiplicity of small groups from a wide variety of cultural and political positions campaigning on environmental issues in the early 1970s are explored by Derek Wall in his history of the anti-roads movement in Britain. Typical of these campaigns were events organized by Commitment, a Young Liberals group, with the aim of demanding free public transport and car-free streets. In December 1971 Victor Anderson organized a 'reclaim the streets' action that attempted to block Oxford Street in London, with *Peace News* proclaiming it to be 'the greatest street party London had ever seen'. There was a second action in the spring of 1973 held in Piccadilly Circus. Victor Anderson recalled in interview with Derek Wall:

> These actions drew upon emerging networks of green activists and countercultur-
> alists; we had links with people like the Dwarves, [and] a magazine called *Street Farmer*. I mean, we didn't have a lot to do with them but that was all part of what was going on, the *Peace News* thing… and then there was local independent envi-
> ronmental action groups who organized part of the demonstration… There was PEST – Planet Earth Survival Team – somewhere in north London. There were people responding to the issues, setting things up on their own. This is before Friends of the Earth.

Other events of the time included a month of action (from mid-February 1972) by Manchester Non-Violent Action Group to promote free public transport, whilst the Dwarfs along with the Young Liberals, held an anti-car march along the route of the M12. The diversity of groups was acknowledged by contemporary observers. In his chapter in *Campaigning for the Environment* (1974), Roy Gregory commented on the enormous range of interests and attitudes 'united by broad desire to maintain or improve quality of environment'. He remarked that 'it is not easy to accept that the Oxford Street Action Committee, the Dwarves, the Street Farmers, and groups firmly embedded in the hippy world are part and parcel of a movement which also includes bodies such as the Civic Trust, the CPRE and the National Trust'. It was precisely this combination of 'the hippy world', traditional organizations such as the RSPB and the RSPCA, and newer ones such as the Conservation Corps and ConSoc that formed the universe of the early eco-activist. These new groups, having a large branch network, were often the first society for young people interested in environmental issues. However because of its staid image and methods, young people tended to drift away from ConSoc to more radical groups such as the Young Liberals, and later FoE. Nevertheless ConSoc played an important role in the development of the emerging environmental movement of the 1970s and as Wall commented: 'The society acted as a particularly significant bridge between middle-class conservation groups such as the civic societies and more radical green campaigners like the Dwarves'.

The formation of FoE UK

The use of friendship networks (or what sociologists call 'social networks') to recruit activists to work on environmental projects is well illustrated by the setting up of a national branch of FoE in Britain in 1970. FoE had been started by David Brower in 1969, when he quit the Sierra Club in California, which resisted his attempts to adopt a more radical and confrontational approach to what he saw as 'an emerging ecological crisis'.[59] Brower realized that any action to halt a global problem would have to be tackled at an international level and in 1970 he travelled to Europe to 'spread the word' and encourage the formation of branches of FoE in France, Sweden and Britain.

He started in France with help from two Americans, Amory Lovins, a physics student, and Edwin Mathews, a Paris-based lawyer. Their efforts led to the founding of Les Amis de la Terre in France by Brice Lalonde – a young economist but latter to serve as Minister of the Environment from 1991 until 1992 under the presidency of Francois Mitterand. While on holiday in Ireland in August 1970, Mathews met Barclay Inglis, a retired Scottish businessman. They talked about the possibility of forming a British FoE group and Inglis decided to organize a special dinner in London to recruit potential activists. The dinner at the exclusive Travellers' Club in The Strand was attended by Brower and fourteen guests invited by Inglis.

Amongst them were Graham Searle and Jonathan Holliman. Searle was vice-president of the National Union of Students (NUS) and chairman of the NUS committee on the environment and had given a speech at the 'Countryside in 1970' conference, whilst Jonathan Holliman was a young writer and campaigner then working for the International Youth Federation for the Environment. A decision was made to form FoE UK, not as a charity but as a company limited by guarantee and owned by its seven original members. This was to allow it to campaign for political change in order to protect the environment, something that was not allowed for under the Charities Act. As Weston commented: 'FOE was designed to be a small, highly specialised, pressure group which could concentrate its efforts on forcing changes to the policies of national government. Unlike the conservationist groups, FOE did not seek or particularly want thousands of members. It wanted to be small, centralised and dynamic'. To this aim the seven members of the company placed the running of the organization into the hands of three people, Barclay Inglis as chairman, Richard Sandbrook as company secretary, and Graham Searle as the organization's first director.[60] In October 1970 the new company, Friends of the Earth UK Ltd., with its three full-time staff, Searle, Holliman and Janet Whelan, moved into an office provided by Ballantine Books at 8 King Street, London. In order to raise funds they started work on British editions of the American environmental paperbacks, among them *The Environmental Handbook* and *The Consumers Guide to the Protection of the Environment*.

FoE's bottle dump: gesture as theatre

There was much discussion within FoE during early 1971 over its initial campaign. It was narrowed down to three issues: opposition to aspects of the fur trade, to the proposal by Rio Tinto Zinc (RTZ) to mine in Snowdonia National Park, and to the decision by Cadbury-Schweppes to switch from returnable to non-returnable bottles. But it was more by luck than design that FoE's first action, the return of bottles to Cadbury-Schweppes' offices on Saturday 8 May 1971, achieved phenomenal publicity and launched FoE into the public's attention. As Weston remarked: 'The bottle dump event was really a media coup for FOE. That style of political activity had not been seen in Britain before and was, until then, more associated with the American system of pressure group politics'.

This style of action, protest as theatre, was typical of Californian countercultural activities pioneered by the San Francisco group, the Diggers, and had been used by Ecology Action – the first hippie environmental group founded in Berkeley in January 1968. A similar group was founded in Boston in 1969, and amongst their planned activities (according to their November 1969 newsletter) was 'a recycling project in which they would try to persuade beverage companies to re-use the glass and metals in their bottles and cans' and their Garbage Group 'was planning a "dump-in" possibly on the steps of the State House'.

The strong influence of American student politics on the activities of FoE in the UK is acknowledged by Philip Lowe and Jane Goyder in their analysis of FoE's style and strategy. They stated that 'Friends of the Earth's roots are American, not British' and noted that this new breed of environmentalism was radical not only in its broad field of concern but also in its campaigning methods. They commented: 'FoE's approach was more open, media-orientated and confrontational, again betraying its American origins. It also incorporated notions of participatory democracy and forms of direct political protest, such as boycotts, sit-ins, marches and demonstrations, borrowed from the student movement'.

Eco-activists needed plenty of free time and few commitments and thus it is not surprising that students were attracted to FoE in the early 1970s. Also there was disillusionment with the other great causes of the 1960s. Student protest had degenerated into violent street demonstrations and often harsh reprisals from the police (especially in the United States) and the counterculture had become aimless with its unintellectual diet of 'sex, drugs and rock and roll'. Thus a movement that combined the intellectualism of student politics with hippie style and humour proved very popular with non-Marxist students, especially Young Liberals. As Walt Patterson remarked, in his chapter in the book *The Environmental Crisis*, 'FoE's Schweppes "demo" gave the word a novel slant: instead of a bitter, even ugly confrontation, it was witty and engaging'. Or as Graham Searle said in a 1974 interview with Patrick Rivers, 'I hope that one of the things we've added [to the environmental movement] is a bit of humour'.

The result of FoE's first bottle 'demo' was extensive media publicity and with it, growing membership and influence that was to make it the leading environmental group in Britain by the mid-1970s. However in terms of numbers it remained smaller than ConSoc until the late 1970s – having only 1,000 members (termed 'supporters') by the end of 1971, just 5,000 in 1976 and 16,000 by 1979. As the 1970s progressed FoE became a national organization by harnessing and expanding its local groups which, with a winning combination of action, hippy style and moderate politics, attracted young activists from existing groups, such as ConSoc and the Dwarfs, both of which went into decline and have now been largely forgotten.

Early anti-nuclear activism

Early anti-nuclear activism reflected the style and tactics of existing conservation groups, mainly middle-class and middle-aged. The earliest protests, such as at Bradwell, involved letter writing and conventional political lobbying. The only demonstration was at Trawsfynydd, and that was by locals in favour of the nuclear reactor! The protest at Stourport in 1970 again followed conventional patterns of lobbying in the press and of government, but this time there was the founding of a group, the Anti-Powerport Station, to oppose the reactor. Their rhetoric was also more militant, with calls for a register of objectors and of 'militant anti-nuclear power station rallies' with further talk of raising money to employ a barrister to state their case at a public inquiry.

In the early 1970s – unlike the USA, France or Germany – there was no mass anti-nuclear campaign in Britain. Partly this was because no new stations were being proposed, and partly because there was little public discussion of nuclear affairs. The first national campaign was started by ConSoc in 1974, but its conventional approach attracted little support from young environmentalists. By late 1975 Beryl Kemp, its organizer, reported that she was becoming disillusioned with the lack of progress through the 'constitutional' approach and wanted to use a new form of action, 'non-violent resistance action', which seemed so successful overseas.

However it was not the tactics that were wrong, but the dullness of the anti-nuclear power campaign run by ConSoc. The campaign did not attract many followers until movement intellectuals, such as those around the magazine *Undercurrents*, were able to establish an 'Other' against which a movement could react. It was this articulation of an anti-nuclear identity that *Undercurrents* achieved with its first issue on nuclear power in early 1975. This publicity for an anti-nuclear campaign was then harnessed by FoE, who were able to launch their own campaign in May 1975. This was a typical piece of FoE street theatre for the media – a birthday party and cake with candles, to mark the tenth anniversary of the start of work on the still unfinished Dungeness B nuclear power station.[61] The birthday party attracted little public attention but a phrase FoE used in their tabloid style paper *Nuclear Times* published at the same time was unwittingly – like FoE's initial bottle dump in 1971 – to achieve intense media coverage and launch anti-nuclear concern to the top of the environmental agenda. The next chapter seeks to explain what it was about nuclear power that roused such concern and opposition.

Chapter 14

Recruiting anti-nuke activists

W hy do people become anti-nuclear power activists? How are they recruited? Do they start off as environmentalists, and then become anti-nukes? Why do only a small minority of those with an anti-nuclear attitude decide to protest. These questions are a subject of much fascination to academics. There are diverse theories on 'new social movements' and 'activist identity' used to explain why people become activists. But as Derek Wall commented, these theories that imply that 'identity grows from a single factor such as class or upbringing or whatever, are inadequate', and that they fail to 'indicate why the majority of surveyed individuals who, due to such traits, can be thus classified as potential activists fail to become involved'. An alternative approach on activist recruitment stresses the importance of 'friendship networks' that encourage uncommitted or disinterested individuals to participate in protest activities. Douglas McAdam in a paper in the *American Journal of Sociology* (1986) gave the following example: 'Imagine… the case of a college student who is urged by his friends to attend a large "anti-nuke" rally on campus. In deciding whether to attend, the potential recruit is likely to weigh the risks of disappointing or losing the respect of his friends against the personal risks of participation. Given the relatively low cost and risks associated with the rally, this hypothetical recruit is likely to attend, even if he is fairly apathetic about the issues in question'.

The use of social networks to recruit activists to a particular cause was noted by Alan Mazur who commented that 'several studies of recent social-protest movements indicate that recruitment often occurs along pre-existing social links, and that people frequently join in an organizational block rather than as isolated individuals'. From his interviews in 1973, Mazur found that about two-thirds of the thirty leaders of the Consolidated National Intervenors – a coalition of more than sixty environmental groups fighting against the nuclear reactor programme on safety grounds – had been in environmental groups. These leaders were usually mature, middle-class liberals who initially had no specific dislike of nuclear power. Their concerns had developed largely through their participation in local environmental groups against the building of a large industrial plant – the nuclear power station – in an area of high scenic value, such as the Californian coastline. Mazur wrote: 'Half of these environmentalists explicitly reported that their concerns had been influenced by other antinuclear people with whom they had come into personal contact. A few nonenvironmentalists were introduced into antinuclear groups through friends... Only a few respondents formed their anti-nuclear alignment completely independently of important social influences'.

Mazur even found that the opinion of academics on nuclear issues was related to the social networks which existed within the academic science community. He concluded: 'The political judgment of scientists are shaped by their social milieus, just as those of laymen'. Polls showed that a large proportion of the one fifth of the population that were always hostile to nuclear power were also opposed to the degradation of the environment and corporate control of society. The left orientation of the opponents of nuclear power reactors could be observed as early as 1956, and Mazur noted that 'liberals were more likely than conservatives to find the antinuclear issue appealing, and were more likely to be recruited by the early liberal activists'. The ideology of collective rather than individual action meant there was a long tradition within the left of individuals banding together to form campaigning groups, which then combined to form a mass movement.

The anti-nuke movement in the US

The anti-nuclear movement in the USA arose partly due to policy decisions made by the nuclear industry, which alienated some of its traditional supporters, but mostly due to external changes in society that turned passive acceptance of nuclear into open hostility for a significant minority of the population. The policy decisions made by the nuclear industry were typical of any maturing technological industry, as bureaucrats take over from scientists and as the profit motive replaces early idealism. As the industry matured and more reactors came on line, the emphasis was now on regulation rather than research, on everyday safety rather than hypothetical accidents, and on consensus rather than critical inquiry. Nuclear scientists were no longer in control of a technology they had created. Those scientists who fell foul of the bureaucratic line, such as John Gofman and Arthur Tamplin, were often eased out of the industry, sometimes to become embittered and highly vocal critics. However the nuclear industry has, since its birth, always had its critics, often ex-members of staff, but their criticism rarely reached the public. Why then did the public suddenly want to listen to nuclear critics from the late 1960s?

In the mid to late 1960s activists in the small anti-nuclear minority in public opinion may have taken the opportunity of the absence of nuclear power articles in magazines as an opportunity to speak out – previously their voices would have been drowned out by nuclear optimists. Also after a dearth of nuclear articles, magazines may have welcomed a new opinion, a fresh approach to the hackneyed subject of the wonderful benefits of the peaceful uses of nuclear energy which filled pages of magazines in the late 1950s, and particularly at a time of national debate over nuclear weapons centering around the Anti-Ballistic Missile (ABM) programme. Ernest Sternglass used a novel approach, in his controversial article 'The Death of All Children' in the mass circulation magazine *Esquire* in September 1969, linking the dangers of radiation from nuclear weapons tests to high rates of infant mortality. He achieved wide publicity for his radiation statistics claims, even though they were rapidly discredited by nuclear scientists who were disgusted by his attacks and who were encouraged by the AEC to refute him. As a contemporary article about him by Philip Boffey, in *Science* in October 1969, remarked: 'Indeed for a man who is so widely regarded as wrong, Sternglass has achieved surprising exposure on the nation's airwaves and in the mass media'. However as Spencer Weart commented:

> While the experts were dismissing Sternglass, ordinary citizens who read his state-ments or saw him on television could not be so sure. He was a physics professor, just like others who were arguing plausibly against the ABM. He was also a good performer, making his points clearly and persuasively. When he and an AEC expert each had a few minutes to present opposing views on a television show, it was impossible to tell who was right.

Boffey asked how Sternglass could achieve such wide exposure for his views when so many scientists believed he was wrong. Part of the answer, Boffey wrote, probably lay in the fact that Sternglass made good press copy – he had a startling theory that related to an important public issue. Furthermore Sternglass 'is in tune with a number of deep public moods – the revulsion against the military, the desire to end contamination of the environment, and the tendency to disbelieve the rosy reports emanating from government agencies'. Scepticism about the impartiality of government agencies was further reinforced by the AEC's treat-ment of two of its employees, John Gofman and Arthur Tamplin, at its Livermore Laboratory in California. Gofman, as an expert on the radiological impact of biologically absorbed plutonium and the leader of a health research programme, had been asked to review Sternglass' *Esquire* article, and had enlisted the aid of Tamplin. But Tamplin's report, defended internally by Gofman, did not please the AEC, even though it did reduce the number of infant deaths a hundred fold from 400,000 to 4,000 – the AEC found the number still too high for its public image. As Weart commented: 'Like any bureaucracy, the AEC disliked doubters within its own ranks, and the pair found themselves increasingly unwelcome at Livermore'.

Gofman, however, was a natural rebel who despised government authority in general, and was happy to fight against the bureaucrats of the AEC over the impact of low-level radiation from nuclear reactors. Challenging the claims that reactors were 'absolutely safe', Gofman and Tamplin toured the nation, giving talks and interviews to warn against materials routinely emitted by reactors, and to promote their book '*Population Control' through Nuclear Pollution* (1970). Gofman and Tamplin built upon the worries that Sternglass had originally provoked, and received generally favourable media coverage especially on TV. On one TV documentary, 'Powers that Be', broadcast on 18 May 1971 in Los Angeles, Weart remarked: 'Viewers were left with the thought that, as narrator Jack Lemmon put it, "nuclear power is not only dirty and undependable… it's about as safe as a closetful of cobras."' Against such anti-nuclear propaganda the AEC was steadily losing 'the battle of images'. It was also losing the scientific debate over threshold doses of radiation and in 1971 the AEC was forced to reduce the maximum permitted dose that a reactor could emit.

Anti-war radicals turn anti-nuke

The criticism of the AEC must be seen in the context of radical student protest against the Vietnam war and anything that smacked of the 'military-industrial complex'. It was an easy step to move from protesting about bombs and missiles to reactors. As public interest in the ABM proposals faded away in late 1969, groups such as the Union of Concerned Scientists (UCS) and the Scientists' Institute for Public Information looked around for a new issue. They found it in local protest to the siting of nuclear power reactors. The UCS was originally a group of Boston-based scientists and students who had come together in 1969 against the ABM and the military uses of academic research and wanted greater democratic control of science and technology. After the ABM protest faded away membership declined but some were drawn into the controversy over a proposed nuclear site near Boston. One such member was Daniel Ford, then an economics graduate student at Harvard, who got involved in research over the safety of the ECCS (emergency core cooling systems). In July 1971 the UCS published a report exposing the inadequacies of the ECCS and the internal dissent. It attracted widespread media coverage.

Hoping to answer this new public criticism, the AEC opened public hearings on ECCS in January 1972. These were planned to last just six weeks but extended, with interruptions, to well over a year. Robert Gillette, in a series of four articles in *Science* in 1972, traced the roots of the safety issue 'to problems of the management of the nuclear safety programme, and to an intense discord that has developed between the AEC and its national laboratories'. This discord and the adverse publicity it received over its stifling of internal dissent and its scornful dismissal of external critics fatally weakened the AEC's authority and credibility. Its behaviour only reinforced people's suspicions about it being a powerful and arrogant bureaucracy or what was called an 'atomic establishment'. Even the industry trade journal, *Nucleonics Week*, noted that the hearings 'have opened up a Pandora's box of scientific doubts and bureaucratic heavy-handedness'.

Writing about this era Daniel Ford, in his book *The Cult of the Atom*, observed that the Vietnam war protests had prompted widespread attacks on 'establish-ment' endeavours and that the legal challenges to the nuclear power programme were raised by citizen activists, local and national government organizations, 'public interest' law firms, and ad hoc groups 'opposed to the siting of nuclear power stations in their localities'. One example of a public-interest group was the Scientists' Institute for Public Information in St. Louis, led by Barry Commoner, which in 1971 brought a lawsuit against AEC's breeder reactor programme. Their critique was summarized in the first methodically argued and widely read anti-nuke book, *The Careless Atom* (1969), by a Commoner protégé, Sheldon Novick.

Most other organizations that opposed reactors similarly drew much of their strength from individuals and groups who had first fought the AEC over weapons. This crossover from weapons to reactors was common, and as Weart remembered: 'In the coffee houses where students gathered in the early 1970s, bulletin board posters opposing the Vietnam War and the ABM were covered over with notices of meetings to protest reactors'. This mobilization of activist groups is illustrated by the formation of the Consolidated Nuclear Intervenors which challenged the AEC during the ECCS hearings, using the technical evidence of Henry Kendall and Daniel Ford.

The growing anti-nuclear movement was also greatly helped by the willingness of usually pro-nuclear papers, such as the *New York Times*, to give space to stories questioning the competence and credibility of the AEC. This was done using the Freedom of Information Act to force the AEC to disclose secret documents, which were then passed onto friendly journalists. David Burnham, a *New York Times* reporter, who covered the campaign on the ECCS, remembered: 'Henry Kendall and Dan Ford came up with a brilliant way of using the media in their nuclear safety work. They knew that a reporter... could not directly challenge the assertions of organized physicists or engineers that the nation's nuclear power program was a dream come true. But a reporter could write about the administrative problems of the nuclear program, providing there was documentary evidence'.

Thus the anti-nuclear critics saw the weak point of the nuclear programme as its administrative, not its technical, record and knew this would appeal to the 'muck-raking' tradition of US journalism. As Daniel Ford put it: 'Having regarded the nuclear programme uncritically for two decades, the news media detected the scent of a potential scandal, which in the normal course of events it would be likely to dramatize and to build into an even bigger controversy'. As David Burnham remarked: 'These documents gave me – as a reporter for the *New York Times* – the hook that was required to persuade the editors that there might be a problem with nuclear energy. Since the *Times* has always been a true-blue believer in science, technology, and progress, Henry and Dan's achievement – finding the evidence that persuaded the paper to print stories questioning nuclear power – was a prodigious feat'.[62]

By the early 1970s the AEC was worried by the increasingly successful tactics of the vocal minority who were opposing nuclear reactor construction, and the negative tone of much nuclear power coverage in the mass media. The fiasco of the ECCS hearing for the AEC during 1972 deepened its crisis of confidence, and encouraged its internal critics to speak out. As Cohn remarked: 'The long public airing of the ECCS controversy helped legitimize the private doubts of many technocrats within the AEC and nuclear industry and helped induce a flow of leaked documents and protest resignations'. The credibility of the AEC continued to decline with calls by its critics for its dismemberment. This was achieved in 1974, when Congress split the promotion and regulation of nuclear technology into two agencies, and also stripped the powerful Joint Committee on Atomic Energy of its unique powers. This was a major victory for the anti-nukes and nuclear power lost its strongest institutional supporter.

This success for the anti-nuke movement also attracted professional activists, those who saw the opportunity for their organizations to take advantage of an increasingly popular issue.[63] Typical of these was Ralph Nader, a crusader against the hidden evils of industry and government, and in 1970 more concerned with air pollution from oil and coal than hazards from nuclear power. But by late 1972, after the publicity over the warnings on radiation by Gofman and Tamplin and the AEC machinations at the ECCS hearings, he had become convinced that in nuclear power the public faced another official 'cover-up' of hazards. So he convened a national conference in late 1974 that brought together more than a thousand anti-nuke campaigners. According to a journalist writing in *The Nation* in January 1985 those attending 'had the grim flavour and messianic fervor of the movement to end the war in Vietnam'.

UK imports US expertise

Anti-nuclear arguments in the United States were rapidly transmitted to Britain by the media, and used in British debates over the safety of nuclear power. Leslie and Shaw, consultants working for Worcestershire County Council in its assessment of the proposed reactor at Stourport in 1970, used the work of Gofman and Tamplin. They were keen however to distance themselves from the claims of Sternglass: 'Gofman and Tamplin are responsible people... and should not be confused with Professor Sternglass, who recently got a lot of publicity for his views that bomb tests had killed hundreds of thousands of children'. The influence of American anti-nuclear works was also evident in remarks by Dr. Spiers, another consultant to the council, when he said, most probably referring to Sternglass' claim: 'I believe Windscale-type accidents are unlikely or impossible with the present closed-circuit type of reactor, but I imagine some kind of accidental escape is conceivable and I imagine the Council to be concerned about this, particularly in view of the recently renewed press and TV discussions on radiation dangers'.

The work of Gofman and Tamplin on safety had been extensively used in a BBC TV documentary 'A Question of Survival' broadcast on 14 December 1971. This broadcast, later criticized in the technical press, is one of the key events to which Pocock attributed the start in 1972 of 'well-organized and well-publicized opposition to plans for the expansion of nuclear power in Great Britain'. Pocock is certainly wrong here, an effective opposition did not emerge until 1975. However movement intellectuals, such as Walt Patterson and Peter Bunyard, had been articulating reasons for opposition to nuclear power in their magazines *Your Environment* and *The Ecologist* since 1970. Patterson was in a unique position to do so and to import all the US debate, as he had reported on the 1972 ECCS hearings. Opposition to nuclear power amongst movement intellectuals was fuelled not so much by 'nuclear fear', as by the growing realization that nuclear power was not going to be the utopian energy source – the answer to the world's energy problems – that pro-nuclear intellectuals like Alvin Weinberg had promoted. Thus John Davoll, in his review of nuclear power in the ConSoc 1975 Annual Report, not only highlighted the problems of the 'rapidly growing stockpile of long-lived, indestructible and highly toxic radioactive wastes' but also argued that 'detailed proposals for a world nuclear economy, by Alvin Weinberg and others... only undermine forcibly how completely it commits humanity to maintaining a complex and hazardous system... Nor does nuclear power, on careful examination, show promise of being cheap or being able to mitigate the impending confrontation between rich and poor countries'.

Why anti-nuke?

What converts the individual to protest about nuclear power, rather than a myriad of other worthwhile issues? Sometimes it comes as an almost religious conversion; the sudden association of nuclear reactors with nuclear weapons, as a life-threatening technology. One example is the case of David Pesonen, a leader of the group opposing the building of a reactor at Bodega Bay in California, who had at first only opposed the reactor as an industrial object desecrating the shore. According to Sheldon Novick, a chronicler of the campaign against nuclear power in California in his *The Electric War* (1976), Pesonen underwent a remarkable experience whilst meditating on his fight: 'It was a beautiful evening, a touch of fog,' Pesonen recalled. 'I had a feeling of the enormousness of what we were fighting; that it was anti-life'. It struck him that nuclear power was 'the ultimate brutality, short of nuclear weapons'. From then onwards he became a vigorous campaigner against all reactors.

Other times it is an aesthetic protest at the mutilation of the landscape, which draws on moral outrage over the degradation of the countryside dating back to childhood days. One such case is that of Langdon Winner, who in the final chapter of his book *The Whale and the Reactor* musing on the cultural and political significance of the Diablo Canyon reactor, wrote: 'To put the matter bluntly, in that place, on that beach, against those rocks, mountains, sands and seas the power plant at Diablo Canyon is simply a hideous mistake. It is out of place, out of proportion, out of reason. It stands as a permanent insult to its natural and cultural surroundings'.

Against the 'Other'

However, by the mid-1970s many leading anti-nuclear intellectuals opposed nuclear power not on environmental or ethical grounds, but because it was promoted by arrogant and secretive institutions. Peter Chapman, in an interview with the author in 1998, remarked: 'What got me was the stupidity and arrogance of the industry. It was the people not the technology I was opposed to. It was the scientific stupidity of their claims it was safe, that there was no problems, such as waste disposal. They had no basis for such claims. Gofman really impressed me, his claims that there was no such thing as a safe level of radiation'.

Gerald Leach, an observer of the nuclear scene in the 1970s, attributed the motivation behind many of the campaigners to dislike of large-scale institutions, and the threat of the 'nuclear state'. Nuclear power, he believed, invited attack as it was 'an attractive target for anti-authoritarians' being arrogant, secretive, and insensitive. This was true for Robin Grove-White, an early environmental activist, who was motivated by his dislike of the 'nuclear state' and its threat to civil liberties. What irked him was the 'CEGB arrogance of assuming they could do whatever they liked'. Another anti-nuke campaigner, Brian Wynne, became involved because he had no confidence in the CEGB's energy projections or forecasting methods. He believed that they had 'no intellectual substance' and were 'anti-democratic'.

The most vocal and widely publicized criticism of the nuclear industry came from Walt Patterson. It centred more on a critique of the electricity supply industry (principally the CEGB) as a powerful, monolithic, and undemocratic institution than on environmental issues of nuclear power. According to Patterson, in an interview with the author in 1996, political opposition to the CEGB coalesced in the anti-nuclear movement because nuclear power was 'a big, easy and unifying target'. He believed that nuclear power was vulnerable because it was 'arrogant and powerful', and could be overthrown in a 'David versus Goliath struggle'.

Furthermore he went on: 'The nuclear industry was its own worst enemy. In the 1950s and 1960s there was boundless enthusiasm for it, the public swallowed every extravagant claim they made, which all proved wrong. Nuclear was given a start no other technology has enjoyed'. He believed that nuclear power had built itself up with myths – of man confronting and overcoming nature, of exploring the deepest secrets of nature, and of nuclear physics as so esoteric and full of powerful secrets that it was fit only for the Weinberg idea of an elite 'nuclear priesthood'. Patterson aimed to debunk this mysticism with his book *Nuclear Power* and explained:

> The nuclear establishment couldn't be trusted. Up till then it was taken as law that only they could understand nuclear power. My book, *Nuclear Power*, dispelled the aura of mystery, it had a big impact. I argued from the onset that there was nothing complex about nuclear, no more than TV, you didn't have to understand how it worked to have opinions on its use. This changed the tenor of the debate. It demystified the experts, they were then seen as overstating and misstating their case.[64]

Finally Gerald Foley, a more detached observer of the nuclear scene in the 1970s, laid the blame for nuclear's unpopularity on its delusion of nuclear grandeur. In an interview with the author in 1996 he commented: 'I was most unimpressed by the whole nuclear establishment. They were not malevolent but just self-serving as others are, they were well-meaning but had delusions'.

Thus there was a divergence in concerns between local people (and old style conservationists) opposed to nuclear power and the new movement intellectuals. For instance at a rally in 1976 at Barrow residents demanded reassurances about possible hazards associated with the transport of radioactive materials through the town, while FoE expressed greater concern about more remote issues of plutonium security, proliferation implications, and the economic validity of the reprocessing industry.

Chapter 15

The anti-nuclear campaign in the 1970s

The year 1970 can be seen as a divide between the old conservation groups and the new environmental ones. It was 'Countryside Year', marking the end of the decade of conferences on the theme of 'The Countryside in 1970', and in Walt Patterson's view, 'the last gasp of traditional conservation'. It was the year of the founding of the UK branch of Friends of the Earth (FoE) in which Patterson was to play a leading role. It was also the year of the first local British protest for a decade in Britain against the construction of a nuclear power station, and also of increasing media (and environmental press) attention to the nuclear opposition in the USA. By 1970 two influential US anti-nuclear books had reached Britain – Curtis and Hogan's *The Perils of the Peaceful Atom* (1969), and Gofman and Tamplin's *'Population Control' through Nuclear Pollution* (1970).

This chapter examines the building of a national network of organizations opposed to the expansion of nuclear power, and in particular to those groups involved in the Windscale Inquiry. Protest was initially small and scattered, but under the leadership first of the Conservation Society and then of Friends of the Earth, it attracted increasing support from a wide range of groups. In the early 1970s protest may have been inspired by the continual drip of articles in the national press on the dangers of low-level radiation. For instance *The Times* gave front-page coverage on 27 January 1970 to the allegations of Sternglass. Then there was a favourable review of Curtis and Hogan's book on 4 March, followed by anti-nuclear letters to the paper. One such letter was by I. M. McNall (on 14 September 1971) where he outlined the conservationist case on energy (similar to that made by Schumacher in the 1960s), which would be adopted wholesale by later environmentalists. He wrote that it was 'time to stop burning up limited stocks of fossil fuels, to stop producing ever-increasing quantities of nuclear wastes, and to invest in the two sources of energy which are nearly perpetual, clean and free – sunlight and the force of the tides'.

This criticism however resulted in little action, mainly because there was no national group willing to promote the anti-nuclear power message. Interestingly Peter Bunyard claimed in his book *Nuclear Britain* that one of the first anti-nuclear demonstrations in Britain was organized by foreigners. It took place on Easter Monday in London in 1970, and 'could muster only a handful of people of whom nearly all were foreigners. The demonstration then merged with the thousands of CND marchers... Many of the CND marchers were surprised at the demonstration against nuclear reactors; they did not consider the connection between reactors and the production of weapons-grade plutonium to be an important issue'.

Nuclear power in the eco-press

This demonstration and press coverage of the dangers of radiation from nuclear waste may have inspired environmental writers. For in mid-1970 there were three articles published critical of nuclear power, concentrating on the dangers of waste disposal. In May John Davy published an article 'Atomic Waste' in *Span* (the Soil Association journal).[65] Then in June 1970 Walt Patterson published his first nuclear article entitled 'Odourless, Tasteless... and Dangerous: Hazards of Radioactive Waste' in *Your Environment*. Finally in the July (the first) edition of *The Ecologist* Peter Bunyard published his article 'Is there a peaceful atom?'

Both Walt Patterson and Peter Bunyard became prolific and well-known writers on nuclear issues. Patterson, then a commentator on UK environmental affairs for the US journal *Environment*, became editor of the UK journal *Your Environment* in 1971. In 1972 after its demise he joined FoE. Bunyard subsequently joined the editorial board of *The Ecologist*, and regularly provided coverage and book reviews for that magazine on nuclear issues throughout the 1970s. He was the first writer in the UK eco-press to report on nuclear opposition in Europe – the protests at Fessenheim, in Germany, on Easter Monday 1971 – and on Scotland's energy and nuclear problems.

Initially the eco-press concentrated their criticism on the American LWRs – the British AGR reactors overall received a relatively favourable press. For instance Patterson writing in *Environment* in December 1972 praised the AGRs as 'more efficient' and 'more environmentally sound'. However wider anti-nuclear arguments were steadily introduced (from the USA), including the risks of low-level radiation, the link between civil and military uses, and the dangers of nuclear terrorism. Then on 14 December 1971 came a BBC documentary 'A Question of

Survival' with its second part on radiation – a reviewer said that its message (echoing Tamplin) was that there was 'no safe dose'. By the end of 1972 Patterson noted with satisfaction, in his second nuclear article in *Environment*, that newspaper reports, television programmes and public discussions were at last beginning to take a more critical attitude towards nuclear power than had hitherto been the case in Britain.

However, unlike the USA, France or Germany there was no mass campaign. Partly this was because no new stations were being proposed, and partly because there was little public discussion of nuclear affairs. The old style conservationists and new left were generally in favour of nuclear power, while most environmentalists were concerned with issues of global doom – the *Limits to Growth* and *Blueprint for Survival* debate. Few had thought about nuclear issues. As Roger Williams in his book *The Nuclear Power Decisions* observed: 'Opposition to nuclear power had increased throughout the seventies in all the liberal democracies, drawing strength from a substantial measure of international co-ordination, but the movement in Britain remained distinctly muted until the Windscale issue in 1976-7'.

Harold Bolter, BNFL's director of information, in his highly revealing book *Inside Sellafield*, placed the start of the anti-nuclear campaign to 1976 with the publication of the Flowers Report.[66] Criticism of nuclear power in this report, he maintained, gave the campaign its early impetus and credibility and meant that the opposition had to be taken seriously. 'The Royal Commission's views could not be shrugged off as the fevered ravings of people with vivid imaginations and suspect political motives – the nuclear industry's customary response to opposition in those days'. Patterson, in his 1979 article, also agreed on the importance of the Flowers Report in 'confirming for the first time in semi-official terms that some of the main points raised by "environmentalists" warranted genuine concern' and that the political credibility of criticism of civilian nuclear power was immeasurably enhanced.

ConSoc debate and early action

The anti-nuclear campaign in the 1970s had its origins in the ConSoc debate over nuclear power that followed the reprint of a BBC Radio 3 talk on the Stockholm Conference by Jon Tinker (given on 27 June 1972). The text, published in *Conservation News* in September, said that the Stockholm Conference had 'totally failed to come to grips with nuclear power. At the moment, only a handful of countries have atomic power stations. In less than thirty years' time, half the world's electricity will come from the atom. By then, two new 500 megawatt nuclear power plants... will be opening every day'. Tinker then described the problem of radioactive waste. He said there was no safe way of disposing of it, and that 'this devil's broth can never be released, for it contains more radio-activity than a dozen H-bombs... What sort of environmental legacy is that for our children and grand-children?' He then warned: 'The trouble is, by the time we find out if the long-term waste problem is insoluble or not, thirty or so of the world's leading nations are going to be hooked on nuclear power, unable to give it up without crippling their electricity supply'. He remarked that at Stockholm the hazards of nuclear power were not on the UN agenda, and hence not discussed officially, but were discussed at many unofficial forums and conferences.

Professor Fremlin, who was active in ConSoc in promoting the nuclear cause and gave talks to local branches, came to the defence of nuclear power in the ConSoc newsletter.[67] He wrote (January 1973) that Tinker has made 'an absurd and incorrect statement that the "devil's broth" contains more radioactivity than a dozen H-bombs'. Fremlin pointed out that the burning of fossil fuels was causing the death of something in the region of tens of thousands of people a year, while the danger of nuclear power was hundreds to thousands times less than that of other methods of power production. In the April edition, John Davoll, the director of ConSoc, said that Fremlin was 'to put it mildly, disingenuous', and that 'not everyone shares Professor Fremlin's optimism about the ability of humanity to manage vast amounts of nuclear wastes safely for millennia'. Davoll's lack of confidence in nuclear experts is apparent in his review of Patterson's first book, *Nuclear Reactors*, in the October 1973 issue where he said: 'Nuclear power is only part of an even more intimidating nexus of problems related to the use and generation of energy... Can any society so dependent on the advice of experts... have any hope of arriving at informed and democratic decisions on the most vital questions that face it?' Davoll again warned, in ConSoc's 1974 *Annual Report*, on the dangers of seeking reliance on increased energy supplies: 'Even more seriously, the commitment to nuclear energy is gathering momentum, and may well soon become almost irreversible; the nuclear waste it will produce represents industrial society's ultimate gesture of indifference to the future'.

ConSoc nuclear policy

On 15 December 1973, ConSoc council accepted an Energy Working Paper (by David Corry, its convener) that expressed opposition to the building of any more nuclear power stations until an absolutely safe method of disposing of the long-lived radioactive wastes had been perfected. Soon afterwards, in early 1974, ConSoc issued a 4-page leaflet *Nuclear Power – Salvation or Deathtrap?* which briefly explained the dangers of nuclear power and gave the society's views. As the ConSoc 1974 *Annual Report* stated: 'It is hoped to use this widely in opposing further reliance on nuclear power, and particularly the implicit commitment to the plutonium-based fast breeder reactor'.

Then at the AGM in November 1974, two anti-nuclear policy resolutions were passed. These stated:

a)   calls upon the British Government to halt the building of any more nuclear power stations while there is no absolutely safe method of disposing of long-lived radioactive wastes

b)   urges that the Society... should launch and sustain a campaign to alert the general public and to inform the Government about the dangers to all life, born and unborn, created by nuclear technology; and

c)   calls for an investigation into alternative sources of power.

There was however much internal division within the Society over its anti-nuclear policy, as is evident from the letter pages of *Conservation News* during the next year. Fremlin was the first to oppose the Pinks' campaign (see below) and defend nuclear power in the January 1975 edition but was rebutted by the Pinks in the next issue. The January 1976 edition had four letters on nuclear power – one defending but three opposing, including one by Beryl Kemp, the new convener of the Energy Working Party. Although ConSoc's public campaign was couched in technical language, about reactor choice, energy supply issues and risks of radioactive waste, it was, according to the letter by Beryl Kemp, at its heart about the moral issue of intergenerational equity. 'The moral issue has been the basis of the campaign which Jane Pink and I (and our supporters) have run for the last year or more. Even if the risks were acceptable to us, we consider that we have no right to endanger future generations in order to preserve our own high living standards'.

The debate on nuclear power was further enlivened by interviews in the spring of 1975 in *Conservation News* with two 'establishment' figures, Lords Avebury and Robens, who both supported its continuation. In the March issue Lord Avebury, ConSoc's president, said he was not in favour of the society's policy of abolishing nuclear power and that 'I must confess that I don't go all the way with the Conservation Society in saying that we shouldn't have a nuclear programme at all'. He was in favour of a small programme of steam-generating heavy water reactors, as it was 'intrinsically a safer reactor than the American pressurised water reactor, because it cannot fail catastrophically'. Lord Avebury believed that the main problem with the expansion of nuclear power was 'not the disposal or storage of the wastes but the risk of nuclear terrorism'. He believed that overall 'we should proceed with great caution and not expand our nuclear programme to the extent that we become unduly dependent on it'. The policy should be 'of preserving a base from which the nuclear industry could expand in the future, if thought desirable'.

In the June issue Lord Robens, the ex-chairman of the National Coal Board, was in favour of nuclear power (despite his great opposition during the 1960s – see Chapter 9). In the interview he said: 'As far as nuclear power is concerned, of course we have to go ahead with this. There can be no argument about it'. Robens was aware of nuclear problems, particular environmental and the threat of terrorism, but he believed that these could be solved by the international co-operation of scientists. He concluded that he did not believe 'that you put nuclear power on one side. You recognize that it has certain problems. You identify these problems with great clarity, then seek the solutions for them'. Despite this support for nuclear power from 'establishment' figures a few activists still pressed ahead with their campaign, largely ignored by *Conservation News* until the Windscale Inquiry two years later.

Anti-nuclear campaign

The initial campaign was led from December 1973 by Jane and John Pink of the Merton Branch, with strong support from John Davoll. The Pinks started their campaign with a short piece in *Conservation News* (September 1974) entitled 'Nuclear Power: the dangers from long-lived radioactive wastes'. In it they emphasized the dangers of long-lived radioactive wastes, and outlined their campaign to bring 'the facts' to public attention. There was also a two-part feature by Peter Dickson entitled 'Plutonium Nightmare' on the possibilities of nuclear sabotage and terrorism in the November 1974 and March 1975 issues.

The Pink's campaign consisted firstly of a 'letter of concern' to be signed by prominent people, and linked to this a petition to 'demonstrate to the government that this is a matter of concern to many people'. The 'letter of concern' appeared in *The Guardian* on 7 January 1975, with 43 signatories including E. F. Schumacher, Bishop Hugh Montefiore, and Peter Hain. The letter stated: 'Our nuclear power programme represents a Faustian bargain in which we are jeopardizing the safety of future generations and their environment for our own short-term energy benefits… We consider it is immoral and unwise to pursue a technology which will leave such a dangerous legacy to posterity'. A copy was sent to all MPs, and the Pinks hoped that 'If it got good publicity, it should give impetus to the local campaigns and we hope thereby attract more supporters for the petition'.

Downing Street march

The next ConSoc action was delivery of the petition (as planned) to Downing Street on 22 March 1975. According to the 1975 *Annual Report*:

> Petition forms were distributed to Friends of the Earth groups, Young Liberals and student environmental groups, as well as Conservation Society members and branches, and a final total of 8906 signatures were obtained to the petition... On 22 March 1975, 90 people were present to march to the entrance of Downing Street; the petition was delivered to No 10 by Alison Pink and Irene Coates, accompanied by Professor Tom Kibble, Dr Kit Pedler, Peter Hain and Diana Heeks.

TV, local radio and the national and local press covered the event. An oblique mention of this petition is made the next day in *The Observer*, which gives a full page to the dangers of nuclear waste. However the march had minimal coverage in the environmental press and only a few lines appeared in *Conservation News*. This is not surprising as ConSoc's campaign had hitherto attracted no attention from the environmental press; the first mention was in *Undercurrents* no. 10, just before the delivery of the petition to Downing Street. Nuclear issues had only received prominence in the previous issue of *Undercurrents* in early 1975, which in a special issue entitled 'Nuclear power – the devil we don't know' devoted sixteen pages to nuclear power. In an article entitled 'Methinks we do protest too little' *Undercurrents* bemoaned the lack of nuclear opposition in this country compared to others and remarked: 'So how much longer will we in Britain go on swallowing our nuclear medicine without a murmur?'

Furthermore at that time local branches of FoE had little interest in nuclear issues, so had little representation on the march. In a news piece entitled 'Where have all the FOE'rs Gone?' *Undercurrents* no. 11 (Spring 1975) reported: 'The march to Downing Street on March 22nd... was supported by only 90 people, mostly from ConSoc. FOE'rs, though supposedly the more radical group, were thin on the ground. The Camden FOE Group, for example... elected to spend the morning shifting the tons of (now valueless) waste paper they had been diligently collecting. And *they* are supposed to be one of the most active and radical groups. A fine sense of priorities'.

A meeting after the march to decide on future strategy ended inconclusively. *Undercurrents* reported that ConSoc and FOE agreed to go their separate ways, with the former preferring the constitutional approach – letters to MPs, and so on. Jane Pink, commenting in *The Ecologist* in December on the effectiveness of the petition, remarked: 'It is always difficult to assess the effects of this type of action, but there is no doubt that the Department of Energy, MPs and others are now more ready to acknowledge that there are legitimate grounds for concern'. Following this petition ConSoc was invited to give evidence to the Royal Commission on Environmental Protection (RCEP) on the hazards of nuclear power. This was done at a meeting held on 11 April 1975, attended by Amory Lovins and Walt Patterson from FoE, together with Fritz Schumacher and John Davoll.

The extent of support for the ConSoc campaign is hard to judge, as it was only mentioned a few times in ConSoc publications and very little in other environmental magazines. In the 1974 *Annual Report*, Davoll mentioned briefly the campaign on nuclear power – that much effort had gone into it and 'most of this has been devoted to increasing public knowledge about the implications, principally because Jane and John Pink of our Merton Branch were willing to initiate and manage a substantial campaign'. In the January 1975 newsletter the Pinks however reported 'that there had been considerable support from Members and others outside the Society for the project to publicize these dangers'. Beryl Kemp took over the campaign from the Pinks but by late 1975 was reporting to ConSoc council on the lack of progress and that she had spent most of 1975 in 'endeavouring to stop the nuclear power programme (with little result I fear)'. Disillusioned with ConSoc's 'constitutional' approach, she intended to 'step up the pressure in 1976 in the form of non-violent resistance action, since other democratic methods seem to have failed'. Her calls for 'direct action' while popular with some anti-nuclear campaigners would undoubtedly have been opposed by John Davoll, and also by FoE.

Early FoE attitudes

As Tom Burke remarked in an interview in 1996 it is now forgotten that in the early 1970s FoE was predominantly in favour of nuclear power. During its first two years FoE 'broadly saw nuclear favourably' but by 1973 'we saw the idiocy of nuclear, being influenced by Amory Lovins and Walt Patterson, who understood the technical arguments and convinced us'. Patterson was well aware of the nuclear (and the PWR) issues in the USA, having reported for the US FoE newsletter *Eco* on nuclear safety hearings in Washington in August 1972. He also knew there was a difference between the USA and the UK in public knowledge and concern, remarking in a 1996 interview that: 'In the late 60s and early 1970s there never was much public concern in the UK compared to the US, but there was an undercurrent of concern among the well-informed'.

Whether FoE was against all nuclear power or just some types of reactors is still unclear. Walt Patterson however still denies in interviews that FoE were part 'of an anti-nuclear campaign or movement' and he dislikes being labelled 'anti-nuclear'. Patterson has always been careful to give the impression that FoE was not anti-nuclear, and claimed he was only against one reactor type, the PWR, largely on safety grounds. Neither has he been concerned about the hazards of low-level radioactivity. He was also dubious about concentrating on radiation issues, unlike the early nuclear opposition in the USA who cited such books as *Perils of the Peaceful Atom* (Curtis and Hogan 1969) which he considered lunatically over the top, though he respected the work of Gofman and Tamplin.

In the early 1970s there was division in the UK nuclear establishment about whether to abandon British technology and go for the American PWR (as the CEGB wished), or press on with a new AEA design (the steam generating heavy water reactor).[68] Here was an opportunity for nuclear critics to influence policy, and FoE seized their chance at the 1973-74 select committee (on science and technology) hearings on nuclear power. Patterson was able to exploit anti-US feeling amongst the British nuclear establishment and as he remarked over two decades later:

> Lovins was all for going guns blazing in an attack on nuclear power, but I said no, there was no public concern or constituency against nuclear power. I knew the British nuclear establishment hated the US nuclear one, ever since the MacMahon Act. Hence the PWR was absolute anathema to them. I argued that we should focus on the problems of US reactors. Reluctantly Lovins agreed and we ran the campaign on this. Within a few weeks the UK nuclear establishment was up in arms.

FoE's submission highlighted the safety concerns with the PWR, and drew on the extensive critiques done by the American anti-nuclear movement. As Roger Williams remarked: 'FoE's principal initiative in 1974 was a memorandum of over 30 pages and with some 80 footnotes which two of its members, Amory Lovins and Walter Patterson, submitted to the Science Committee. Lovins was to become one of nuclear power's chief international opponents, while Patterson was effectively to lead the British campaign for FoE'. Thus FoE established themselves over the period 1974-75 as 'reputable' nuclear critics and a source of dissenting technical expertise. At first the principal lines of FoE's opposition were geopolitical and economic, not environmental, with the early issue being the misuse of resources. As Tom Burke remembered: 'The development of arguments over the period 1973 to 1976 was first the accident risk of PWR, then rate and magnitude of building 40 GW of nuclear, then safety, then resources, then economics, then electricity generation, and finally soft energy paths. After the Flowers Report FoE added work on proliferation issues, and opposition to fuel cycle and then reprocessing'. After the select committee hearings the PWR was no longer in the news and there was little to campaign against. The only public inquiry was in June 1974 over a proposed station at Torness in Scotland, at which Patterson gave evidence against, to little avail and publicity.

Lack of campaign

The FoE leadership were perhaps stung by criticism in *Undercurrents* of their lack of support for the ConSoc anti-nuclear campaign. As *Undercurrents* no. 11 (Spring 1975) reported: 'FOE are continuing their low key campaign. They are concentrating on building up a solid base of well-informed opinion in the country as their first priority; they are trying to stir up local opposition at proposed power station sites; and Walt Patterson is writing a new pamphlet – *The Fissile Society*'. Patterson had indeed been trying to interest FoE members in nuclear issues, without much success. The nuclear debate was in the elite media but not at the grassroots. FoE's defence of its lack of nuclear campaign is intriguing, and perhaps disingenuous, in the light of its subsequent activities. *Undercurrents* reported: 'FOE pointed out that they couldn't orchestrate an anti-nuclear campaign from Poland St' [their London (head) office]. This *Undercurrents* believed: 'This is certainly true: they have a highly anarchic consti-tution which allows each local group to decide for itself what issues to campaign on'.

The relationship of FoE local groups to the head office was certainly complex and a source of much conflict, but while the constitution may have been anarchic (a contradiction in terms?), the power always lay, as local groups frequently complained, with the head office. As Tom Burke reminisced to Robert Lamb, in the book *Promising The Earth*, on the London office in those days: 'These were mythical figures, magical people we were in awe of'. Furthermore he saw his job, as the first local groups co-coordinator, as mobilizing the groups to support head office campaigns. Burke became director of FoE in 1975, and with what Lamb calls his 'robust management style' was able to focus and direct FoE's campaigns towards what he termed 'achievable objectives' instead of 'dithering over the directions its campaigns should go'. The lack of focus and radicalism in FoE in early 1975 was clear to *Undercurrents* (no. 11) when it said: 'There is a widespread realization that FOE is about something more important than collecting wine bottles and wastepaper, but this feeling has yet to crystallize into a coherent consensus. Hopefully, when it does, the results may be quite startling'. It then urged 'disaffected radicals of all kinds to join their local FOE group... and to work to turn it away from garbage collection to more serious tasks'.

Development of a campaign

Commenting on local groups, Tom Burke said that there 'was no groundswell from the grass roots' for an anti-nuclear campaign in 1975 unlike that for whales and transport where they led. They did not know about nuclear issues, and there was much 'ignorance and apathy'. He remembered that: 'John Price[69] and I went for a two week tour of local groups in 1974 telling them of nuclear issues. The oil crisis had focused attention on energy; we saw nuclear power as a pivotal issue, the nub of the wider energy debate and societal issues. To us nuclear was the litmus test of social choice. We posed the question: do you want to live in a nuclear society?'

Burke remarked that FoE started to campaign on nuclear power because 'we had the intellectual resources – Amory Lovins, Walt Patterson, John Price, Czech Conroy, Mike Flood, and Peter Harper on the margins. The original thinking was done by Amory Lovins and John Price, they drew in and analysed the material. Walt Patterson was the energy journalist and polemicist, he could interpret and communicate, and had a fine sense of judgment'. So as Burke remembered it, 'the anti-nuclear campaign had the resources, opportunities and interests of these people with a profound understanding of energy, which led them to develop sustainable lifestyles as they were convinced nuclear was the wrong direction. We were paving the way, bringing the message to the public'.

Thus it was a combination of factors that launched FoE's anti-nuclear campaign in mid 1975. A more focused head office, committed and expert staff, more radical local groups and an issue on which to campaign – BNFL's expansion plans at Windscale. FoE's first nuclear campaign event was in May 1975 when it held a tenth 'birthday party' for the still unfinished Dungeness AGR station, attended by many people from the nuclear industry and the media, and for which it published a four-page tabloid called *Nuclear Times*. On its front page was a story about BNFL's plans, which would make Windscale 'one of the world's main radioactive dustbins'. This story attracted little attention as BNFL's plans were well-known to the press and generally supported. It had been public knowledge since late 1974, and *The Observer* had even carried an article entitled 'Swedes Dump Atom Waste on Britain' on 20 April 1975, without creating any public furore. Even *Undercurrents* no. 12 did not publicize this remark, and reviewing *Nuclear Times* in the autumn of 1975 just wrote: 'The excellent first issue contains a guide to the nuclear industry, the nuclear fuel cycle and an account of the planned British reactor programme, plus information on researching and organizing against your local reactor'.

Then, on 21 October 1975, the *Daily Mirror* had a front page story, headlined 'Plan to make Britain World's Nuclear Dustbin' followed in November by another story entitled 'Sign Here for Japan's Atom Junk'. These typical *Mirror*-style 'shock reports' on the proposed Windscale reprocessing plant brought widespread publicity to FoE's nuclear campaign. With increased activist interest FoE held their first 'Campaign Workshops' weekend in November 1975 on nuclear issues and developed a coherent strategy for local activists. They were thus in a good position to assume national leadership of the anti-nuclear campaign which up to then had involved only campaigning by local groups.

Chapter 16

Early local campaigns

Local branches of ConSoc and FoE were involved in the mid-1970s in the formation of local anti-nuclear groups, which campaigned against the nuclear power stations at Heysham, Torness, Sizewell and Hinkley. The largest and best known of these were Half Life (at Heysham) and SCRAM (at Torness). Little is known about the other two, Survival (at Sizewell) and the Nuclear Reactor Vigilantes (at Hinkley). With Windscale front-page news from late 1975, new opposition groups sprang up, and long-established ones, not previously involved with nuclear matters, 'joined the fray'. However the proliferation of groups led to the impossibility of agreeing on a unified body to present the opposition case at the Windscale Inquiry, and complicated fund raising.

Half-Life

The first specifically anti-nuclear organization in the UK, Half Life, was formed in Lancaster in early 1975 to campaign against the nearly completed Heysham AGR. It consisted of activists from Lancaster as well as members of the Whitehaven branch of FoE and the north Lancashire branch of ConSoc. A founder member was Paul Smoker, a long-time anti-nuclear campaigner and peace activist at Lancaster University, who had been almost alone in voicing concern, in articles in the local papers from 1971, about routine releases of radiation from reactors. The ConSoc branch secretary Avril Orlawski wrote as early as May 1974 to *Conservation News* over its nuclear concerns: 'We are to apply, not very optimistically, for representation on the local Nuclear Liaison Committee. We have an AGR power station under construction in the borough'. There was further mention of Half Life under the section 'Public campaign on nuclear power' in the 1975 ConSoc *Annual Report*: 'It is worth mentioning in conclusion that another anti-nuclear group, 'Half Life' was established in North Lancashire to oppose the Heysham nuclear power station proposal; Mrs. Orlawski, the secretary of our branch in the area, took an active part in setting up this group and arranging a "nuclear awareness week" that included two symposia and much effective publicity'.

This 'nuclear awareness week' at the beginning of June 1975, which was reported in *Undercurrents*, was the typical mix of street theatre, public meetings and debates. Half Life described their campaign (in *Undercurrents* no. 12) as 'a multifaceted attack on nuclear power, comprising local and national lobbying, a sustained publicity campaign and, when we have sufficient community support, direct action where necessary. A crucial part of the campaign is to press for alternative energy policies'. During this week of action there was a public meeting attended by about eighty people. Half Life reported that the response of the public was encouraging, and that local councils and MPs were pressing questions about nuclear safety to the 'electronuclear establishment' but the 'CEGB has resolutely boycotted our public debates and refused to answer questions'. Throughout 1975 Half Life continued their campaign against the Heysham station, which was widened to opposition to the proposed Windscale reprocessing plant. A few days after the *Daily Mirror* story, Half Life presented a petition on 25 October to Downing Street against the proposed plant. There were also protests by Half Life and ConSoc at Barrow-in-Furness (the local port for Windscale) against the arrival of spent fuel from Sweden destined for Windscale. These attracted the sympathy of dockside cargo handlers at Barrow who voted (in late 1975) not to unload any more casks of spent fuel bound for Windscale – a decision which was later reversed by one vote. BNFL, disturbed at these stirrings of public dissatisfaction, hastily arranged a meeting at Barrow town hall on 12 December 1975, with a near capacity audience. According to Walt Patterson's report in the journal *Environment* (March 1976):

> Local dignitaries, union representatives and spokesmen for Half Life and Friends of the Earth were invited to join BNFL on the platform to air their differences... The discussion was wide-ranging and intermittently heated. Local residents demanded reassurances about possible hazards associated with the transport of radioactive materials through Barrow, while Friends of the Earth spokesmen expressed more concern about plutonium security, proliferation implications, and the economic validity of the reprocessing industry.

The next day there was a Half Life rally at Barrow where a good-humoured crowd of protestors staged a demonstration through Barrow, carrying banners and talking to bystanders about their reasons for objecting to the nuclear shipments.

Half Life's campaign for a public inquiry

On 1 June 1976 *The Times* reported that Half Life and the local branch of FoE had decided to press for a public inquiry to be set up as soon as BNFL submitted its expansion scheme for planning approval to Cumbria County Council – which it did that month. But the council would not be rushed into taking a quick decision. The activities of the protestors had begun to cause resentment among some local people, particularly trade unionists who worked at the Windscale plant. Many trade unionists did not share Half Life's fears about the environmental and other hazards of Windscale and welcomed the prospect of one to two thousand extra jobs being created in the area, a view supported by the national executive of the General and Municipal Workers Union (GMWU) who, at their annual conference on 8 June 1976, opposed a motion calling on the government to hold a public inquiry. Local trade unionists also started a petition supporting the Windscale expansion plans, to counter an opposing one organized by Edward Acland.[70]

Cumbria County Council, as the main part of their consultation exercise, organized a public meeting at Whitehaven civic hall for the evening of 29 September 1976. It was well attended and was widely covered by both national and local media, with Radio Carlisle broadcasting the whole meeting live. *The Observer* reported (3 October 1976): 'The hall was packed: 800 people, many diverted into an overflow room, stayed through the whole four hours in the heat and discomfort of the dazzling television lights. The battle lines were drawn for a classic confrontation, with environmentalists and residents on one side, and Windscale employees (some in company ties) and their families on the other; safety versus jobs'. Friends of the Earth, Half Life, the Conservation Society, the Town and Country Planning Association and others all made impassioned appeals for the matter to be referred to a public inquiry. Paul Smoker, from Half Life, made clear that their objection was to only one out of the three BNFL proposals – the thermal oxide reprocessing plant (THORP). They were not opposed to the other two – the refurbishment of the Magnox reprocessing facilities, and the development of a method for glassifying high-level wastes.

On 2 November 1976, in spite of the protests expressed at this meeting and objections lodged with the council, Cumbria planning committee decided that in principle they were 'minded to approve' BNFL's application. However they wanted to pass the ultimate decision to Peter Shore, the minister responsible. So they decided to give him every available chance to 'call in' the proposal by declaring that it considered BNFL's proposals to be a 'departure from a fundamental provision of the County development plan'. If however Shore, the Environment secretary, did not call in the application in 21 days, the committee would go ahead and ratify the permission to which it had already in principle agreed. Thus they passed the buck to the national government.

Network for Nuclear Concern

The Network for Nuclear Concern (NNC) was a regional grassroots anti-nuclear group, whose main constituent was Half Life and local Cumbrian FoE groups. It was founded and coordinated by Edward Acland, a district councillor from near Kendal, and one of the local organizers of opposition to THORP. In November 1976 Acland organized a petition calling for a public inquiry, and collected 28,000 Cumbrian signatures in less than ten days. In contrast the Windscale workers collected 18,000 signatures over several weeks demanding a go-ahead for the project. These petitions came at a sensitive time in mid-November 1976, just when Shore was deciding whether to call in the planning proposal. The trade unionists presented theirs first to Shore on 19 November, when a deputation led by Bill Maxwell of the GMWU travelled to London from Windscale. Two days later Acland presented his petition. In an interview with *The Guardian* on 22 November he said there were 'overwhelming reasons for the plans for a large oxide fuel plant to be put before a properly constituted planning commission. The issues are international and national, not local, and they require expertise and a depth of understanding of the implications which are far beyond those available to local authorities'.

By the time of the inquiry NNC had also gained the membership of several hundred individuals, a Mothers Union group, a coalition of Quaker Meetings, and the Northern Friends Peace Board. NNC decided to be represented at the inquiry, and to concentrate on safety and radiation release issues. It decided its best course was to use an amateur advocate, well-versed in the technicalities, and also most importantly open to the group's influence. Brian Wynne volunteered for this position and became the group's spokesman.[71] He had no legal background but learnt rapidly, helped by other barristers at the inquiry including 'opposition' ones, who discussed cases and methods with him in the evenings. At the inquiry NNC concentrated on the issues of environmental discharges and radiobiology, and was keen to show what a local group could achieve in putting up a highly technical case without use of a professional advocate. It therefore rejected a coalition with the TCPA, and the possibility of using Sir Frank Layfield[72] – reputedly the best planning silk in the country – who was TCPA's QC, and instead allied itself with the Political Ecology Research Group (PERG), with Wynne acting as an adviser.

SCRAM

SCRAM – the Scottish Campaign to Resist the Atomic Menace – grew out of a loose umbrella organization formed in November 1975 which included FoE, ConSoc, Edinburgh University Ecology Group, and Science for People. However according to Ian Welsh, a SCRAM activist, it formed as a splinter group from Edinburgh FoE. As one activist said to Welsh: 'FoE wanted us to campaign to save otters – well sod that when the world is being poisoned by plutonium'.

SCRAM's view was that nuclear power was important enough to require a specific campaign focus and could not be just part of a wider environmental remit. In Scotland there had been several years of campaigning on energy issues. In 1973 FoE and ConSoc, with help from Walt Patterson, had been part of an umbrella organization called the North Sea Oil Coalition. Their victory at a public inquiry into building concrete oil platforms at Drumbuie in August 1975 greatly encouraged Patterson and other groups. Commenting on FoE's performance, Robert Lamb wrote that 'in the Drumbuie process it had learned that public inquiries were winnable and that there was scope to play a worthwhile co-enabling role on the side of a local community defending a prized environment against energy Goliaths and the powers-that-be'.

There was then a two-week public inquiry into a proposed AGR at Torness in mid-June 1974. According to Tony Hall in his book *Nuclear Politics*, defeat at the inquiry galvanized the local opposition and when the inquiry report showed that their arguments had been ignored, they set out 'to show that public opinion was on our side'. And, as one of the members put it, 'when public opinion was ignored, direct action was the only tactic left open to us'. In April 1976 SCRAM organized an occupation of the proposed AGR site at Torness. Hall reckons this was 'the first example of direct action to oppose a nuclear power station in Britain'. A hundred protestors camped the weekend at the (then vacant) site. However the style was more 'free festival' than political. According to *Undercurrents* no. 15 who publicized the event: 'There will be music (a pipe band and folk groups) in the barn on the Thorntonloch campsite, an anti-nuclear exhibition, beachcombing and kite flying competitions, and, on the Saturday night, a public meeting in Dunbar'. Another similar 'free festival' was to follow at Torness in May 1978, when a much larger weekend camp on 6 and 7 May was attended by four to five thousand people. SCRAM was also an objector at the Windscale Inquiry, focusing on opposition to nuclear waste disposal in Scotland, and calling for an end to nuclear power. Both these positions brought it into conflict with FoE, particularly over FoE's (reluctant) support for the AEA's drilling programme for disposal sites in Scotland, something that SCRAM was uncompromisingly opposed to.

Survival

Opposition to the proposed SGHWR nuclear station at Sizewell was led by Survival, a Cambridge-based coalition of FoE and ConSoc groups. On 24 April 1976, simultaneously with planned demonstrations at Windscale and Torness, *Undercurrents* no. 15 wrote of *Survival*'s plans to organize at Sizewell 'a picnic with home-brewed beer, street theatre and music… The workers and management have been invited along and *Survival* hope to arrange an impromptu debate on nuclear power on the beach. They are also, as a publicity stunt for the local media, putting on street theatre and dumping Nuclear Dustbins ('Not to be opened until Easter 5,000,000 AD') in the Market Square'.

How this event went is unknown since in the next issue *Undercurrents* only reported on the Windscale demonstration.

Nuclear Reactor Vigilantes

The group Nuclear Reactor Vigilantes was Somerset-based and active in demonstrating against the Hinkley Point Magnox and AGR reactors. Little is known about this group, save that it was organized by Jane Buxton from South Petherton. In *Undercurrents* no. 25 there is a report about a demonstration at the Hinkley Point AGR open day organized by the Somerset branch of FoE and the Nuclear Reactor Vigilantes. They handed out five hundred leaflets to members of the public and Hinkley employees before being escorted off the premises.[73]

A network of groups

Contemporary commentators found it not easy to describe accurately or adequately the extent and the nature of the anti-nuclear movement. Nuclear analysts John Surrey and C. Huggett remarked in 1976 in the journal *Energy Policy*: 'Rather than being a united movement with common interests, the opposition comprises coalitions of different interest groups – some opposing nuclear plants in their locality, some general environmentalist groups wanting tighter controls, and some specifically anti-nuclear'. Another observer, Gavin Weightman, saw a strong 'anarcho-liberationist streak mixed in with conservationist, ecological and politically pink threads'. Writing in *New Society* in December 1979, he said there were four disparate strands:

1.   The broad centre environmentalists, mainly middle-class – like ConSoc and FoE.
2.   The orthodox left-wing – such as the Socialist Workers Party.
3.   Scargill and the Yorkshire miners.
4.   A conglomerate of anarchists and the 'brown bread and sandals brigade', a euphemism for hippies.

There was thus a wide variety of organizations and reasons for opposition and as Weightman concluded the national movement looked 'distinctly disorganized, eco-minded, left-of-centre, and rather lacking in muscle', but there was local group strength such as at Torness. Overall he concluded: 'The people who join such groups would probably oppose a coal-fired power station or an airport on the same site, and many are not perhaps, strictly anti-nuke'.

I shall now outline the main groups that sprang up to oppose the Windscale expansion and which coalesced into a national network. Also involved were individuals from pre-existing amenity, peace and religious groups, who tried to steer their organizations into taking up an anti-nuclear position. For instance the National Federation of Women's Institutes passed a resolution in 1977 urging the government to postpone the building of the FBR until a better way had been found of disposing of the waste – however it did not receive the two-thirds majority needed for the WI to campaign on the resolution.

Nuclear Information Network

On 14 October 1976 a meeting was called by (London) Greenpeace and other organizations to discuss the possibilities of joint action on nuclear matters. It was attended by 29 people from 15 groups with much of the discussion centred upon the differences, mostly of emphasis, between the groups. A second meeting was held on 11 November and it was decided to set up a small working party to sort out the practical details of making a working link without spawning a new organization, and to avoid unnecessary duplication. The name proposed for this organization was 'Central Information on Nuclear Energy Control' (CINEC).

Beryl Kemp from ConSoc was a member of this working party, along with Martin Lowe (Greenpeace), Sheila Oakes (National Peace Council), Sue Boothman and Martin Aitken. They decided to call a public meeting for 11 January 1977 at Friends House, London, and by then the organization was known as the 'Nuclear Information Network' (NIN). This meeting was attended by 36 people from 24 groups – including CANTO, ConSoc, FoE, Greenpeace, Half Life, SCRAM and SERA and various peace groups. The meeting was dominated by the upcoming Windscale Inquiry, and how best to co-ordinate evidence. The ten proposed objectors at the inquiry outlined their cases, with Czech Conroy from FoE saying FoE's main expertise 'lay in economic and technical aspects'.

NIN saw itself as a co-coordinating and information network or exchange, not an organization, it having no staff and a very low subscription. It saw its value in being able to bring together a wide variety of environmental, political and peace groups to campaign on the Windscale issue, with groups still able to retain their diversity of viewpoints. As such it played an important role in building the anti-nuclear movement, and in mobilizing people. By mid 1977 it had 42 groups on its 'membership' list, many of them little more than 'one-person groups', such as Alternative to Nuclear Technology (Mike Filgate), Branscombe 2000 (Sir Kelvin Spencer), Cambridge Energy Group (Tom Pettitt), Campaign against Nuclear Energy (Ken Barker), Nuclear Reactor Vigilantes (Jane Buxton), People for a Non-nuclear World (Renee-Marie Croose-Parry), Safe Energy Petitioners (Jane Pink), and South Yorkshire Nuclear Action Group (Richard Turner).

There was another meeting on 30 April at the TCPA Institute, London to co-ordinate objectors' evidence. The meeting of NIN on 26-27 November 1977 was a two-day post-mortem on the Windscale Inquiry, and in particular FoE's (strained) relationship with the other groups. Held at the TCPA it was attended by 21 groups and 30 people. On the first day (which FoE did not attend) each group present at the inquiry gave lessons learnt from attending the inquiry, in terms of presenting their case, co-coordinating with other groups, and preparing for the expected Fast-Breeder inquiry. On the second day Czech Conroy from FoE attended. His disowning of CANTO a year previously had already caused bad feeling further heightened by the perceived lack of co-operation by FoE with other groups at the inquiry. Conroy was unapologetic, stating that FoE's case was threatened by the disparate number of groups giving evidence and by the duplication of evidence.

CANTO

CANTO (Concerns Against Nuclear Technology Organization) was a London-based group formed in late 1976 by a John Hanson, a film maker, with assistance from Edward Dawson, an employee of ConSoc from 1973 to 1977. It was an ad hoc group with a 'supporters list' very similar to NIN's. Its main activity (and claim to fame) was the organization at short notice of the first anti-nuclear demonstration in Trafalgar Square on Saturday 20 November 1976, three days before the expiry of the deadline to call in the Windscale application. Edward Dawson also organized a letter to *The Times* on Windscale that was published on 5 November 1976, signed by Anthony Woolf, chairman of the Lawyers Ecology Group and other distinguished lawyers. John Hanson is credited with having influential contacts. He attracted 'celebrities' to attend this rally, and then used his contacts to obtain from Sir James Goldsmith a substantial donation to fund objectors at the Windscale Inquiry.

People for a Non-nuclear World

This group was another ad hoc group of individuals fronted by Croose-Parry. On its headed notepaper it had an 'ad-hoc committee' of 21 people including E.

F. Schumacher and James Robertson, as well as activists John Hanson (of CANTO) and Beryl Kemp (of ConSoc). In April 1977 it rapidly collected over two thousand signatures to an open letter to President Carter in support of his nuclear policy and also raised £10,000. Some of this was used to publish the letter, as a full page advertisement, in *The Guardian* on 2 May 1977. The rest was donated to objectors at the Windscale Inquiry – £2,000 was given directly to the Windscale Appeal, and £500 to ConSoc. In a letter to Leonard Taitz, in August 1977, Croose-Parry refers to 'the mass meeting at Trafalgar Square in the Spring, which you suggest'. This was indeed to take place in April 1978, organized by FoE with Arthur Scargill (of Energy 2000) as the star speaker.

PERG

PERG (the Political Ecology Research Group) was based at Oxford and consisted of about a dozen scientists who undertook detailed and 'non-partisan' studies on the social and political issues of nuclear power, particularly proliferation. Peter Taylor, PERG's co-coordinator and member of both FoE and ConSoc, attempted to be an intermediary between all the groups seeking representation at the Windscale Inquiry.[74] At the inquiry Taylor worked with Brian Wynne from NNC on safety and radiation issues – including public acceptability – which received favourable publicity from Ian Breach, the *New Scientist* reporter.

Society for Environmental Improvement

The Society for Environmental Improvement (SEI) was founded by Gerard Morgan-Grenville and was responsible for setting up the National Centre for Alternative Technology near Machynlleth, Wales in late 1973. Morgan-Grenville was interested in taking part in the anti-nuclear campaign. The January 1976 minutes from the ConSoc council revealed that he, in discussions with John Davoll and Edward Dawson, was interested in the possibility of 'non-violent resistance action', similar to that taking place in Europe. He attended many of the NIN meetings and became a founder member of the Green Alliance (see Chapter 20).

Because of the SEI's interest in alternative energy sources and scenarios it became involved in the Windscale Inquiry, producing in June 1977 the report *An Alternative Energy Strategy for the United Kingdom*. This included a foreword by Sir Martin Ryle, who in May had written in *Nature*, expressing his support for alternative energy sources, particularly wind, rather than nuclear power. At the inquiry Justice Parker expressed his interest in hearing about non-nuclear energy options. Thus the SEI presented evidence on non-nuclear energy strategies based on coal and renewables, using John Tyme, their lay advocate.[75] Witnesses included

Arthur Scargill (active in Energy 2000) on the need to double coal output by the end of the century, Stephen Salter on wave energy, and Peter Musgrove on wind energy.

SERA and the left

SERA (the Socialist Environment and Resources Association) mainly through the efforts of Stan Rosenthal, Tony Webb and Dave Elliott were actively involved in raising the nuclear power issue in the Labour movement. Their campaign concentrated on trade union rights and health and safety issues, particularly after the Windscale strike in February 1977. They were also the first to emphasize the employment potential of investing in renewables and conservation instead of nuclear power. SERA was a founder member of NIN.

SERA's opposition to BNFL plans was founded on a libertarian-socialist conviction that the widespread use of nuclear power would require an authoritarian society which would increasingly infringe the rights of working people. SERA presented evidence at the Windscale Inquiry and its witness Dave Elliott argued for alternative energy technologies, as these would enhance the creative and productive skills of workers: in contrast the development of nuclear power and the construction of THORP in particular, were steps on the road to deskilling society and exacerbating structural employment.

The other left group to oppose nuclear power (and to have a general interest in environmental issues) was BSSRS (the British Society for Social Responsibility in Science), an association of radical (and generally marxist) scientists. It viewed nuclear power as a form of economic and social development seriously at odds with the needs of socialist society. Policy on nuclear power varied across the marxist groups. The Socialist Workers Party came out firmly against nuclear power, supporting the Windscale rally in April 1976 and devoting considerable space in its paper *Socialist Worker* to the nuclear issue in 1977. The other major new left group, the International Marxist Group, also came out against nuclear power, whilst other groups, such as the Workers Revolutionary Party, only opposed nuclear power under capitalism. As Dave Elliott concluded in his book *The Politics of Nuclear Power* (1978): 'In general, while many left groups are currently critical of nuclear power as developed by capitalism, there still remains a belief, at least amongst some of the more traditional marxists, that nuclear power could be beneficial under socialism'.

Energy 2000

Energy 2000 was set up by Leonard Taitz, ConSoc chairman from 1977 to 1980, and convener of the transport working party. He modelled it on his previous association with Transport 2000, where he was chairman.[76] He strongly believed in coalitions between conservation bodies and trade unions, and in the need to seek powerful allies in order to be effective. He thus secured the backing of Arthur Scargill, president of the Yorkshire Miners' Union and Energy 2000 was formed at Barnsley in Yorkshire on 2 April 1977. It aimed not only to oppose nuclear power but to press for research into other energy sources, not only coal but also renewables. It presented itself as a cross-party lobby group and was able to attract the support of many MPs interested in energy issues, such as Robin Cook (Labour), David Penhaligon (Liberal), and Nigel Forman (Conservative) as well as outside groups.

Taitz's association with Arthur Scargill was controversial amongst some ConSoc members who viewed Scargill as an 'extreme left-winger'. In a letter in February 1977 to J. H. Goodland – ConSoc chairman in 1968 – Taitz defended his decision to approach Scargill for assistance in the anti-nuclear campaign. 'When I approached Mr. Scargill, it was not in his capacity as a Marxist or extreme left-winger, but as a leader of a powerful trade union, who had expressed anti-nuclear views... [that] come from a powerful personal belief that nuclear energy is dangerous'.

Taitz furthermore wrote: 'On a more fundamental level, it seems to me that the Conservation Society dare not reject support where it is offered, and indeed must seek for allies where it finds them... One of the factors that has prevented the Conservation Society becoming a more powerful organization, is many of the people who would have supported us, have considered it in the past to be too elitist and right-wing'. Writing eight months later to Beryl Kemp, Taitz said after his meeting with Tony Benn and Peter Shore: 'This was an extremely useful occasion and I think firmly illustrates the value of our association with the Yorkshire Miners through Arthur Scargill. It gives us a great deal of muscle, at least while the Labour Party is in power, that we would not otherwise have'. Arthur Scargill brought an air of left-wing radicalism to the anti-nuclear movement. He was one of the main speakers at the Windscale demonstration in London in April 1978, and his call for civil disobedience was well received.

The next chapter describes how these disparate groups attempted to present a common front at the Windscale Inquiry, and how anti-nuclear enthusiasm was no substitute for nuclear expertise, at least in the eyes of the national media.

Chapter 17

The Windscale Inquiry: glory or fiasco?

In late 1975, in the light of media interest and environmentalist concerns over BNFL's proposals, Tony Benn, the new energy secretary, was keen to practise his ideas on 'open government', and thus have a 'great debate' on all aspects of nuclear policy in Britain. At his suggestion BNFL quickly organized a public debate, which took place at Church House, Westminster, on the afternoon of Thursday 15 January 1976. Principal speakers were Con Allday and Peter Mummery from BNFL, opposed by Walt Patterson from FoE and Paul Smoker from Half Life. As Czech Conroy commented in his book *What Choice Windscale?* (1978): 'This was the first time the advocates of nuclear power in Britain had met their opponents head on in a public forum of such importance... The debate itself was rather inconclusive as might have been expected from two such diametrically opposed views meeting for the first time'.

New Scientist (22 January 1976) was less complimentary and wrote that 'nothing could counter the poor showing of the anti-nuclear movement' with only Walt Patterson putting on 'anything like a convincing show'. While Patterson's questioning of the economics and technical basis of the reprocessing business, and his lack of hostility to nuclear power won the audience's respect, Smoker's overtly political case was judged to have carried 'little weight with the nuclear establishment and alienated some of the opponents of the BNFL contract' – presumably Walt Patterson. Harold Bolter, in his insider account from within BNFL, had a more cynical comment on Benn's motivations for this set-piece debate, writing: 'In politics, little is believed to have happened unless it has taken place in London'. He wrote that Paul Smoker accused BNFL and the Department of 'news management' in organizing the discussion, and Bolter commented that 'of course he was right'. Furthermore the debate 'seemed to do the trick as far as

Tony Benn was concerned' for on 12 March 1976 Benn announced that BNFL could take on further reprocessing work for overseas customers and 'that this statement took most of the steam out of the debate'.

The Windscale rally

The first major event of the anti-Windscale protestors was a rally attended by over 600 people at Windscale on Saturday 24 April 1976, organized by FoE and Half Life. FoE chartered a train – dubbed the 'Nuclear Excursion' – to take over four hundred people there and back and this train generated much publicity. As *Undercurrents* no. 15 reported, 'Tom Burke, the pugnacious director of London FOE, hopes these rallies will finally kill off the "soft Schweppes" image that many people still have of FOE'. However FoE was against direct action, preferring debate and discreet lobbying. As *Undercurrents* commented on the plans for this event 'The plant however will not be picketed as originally planned. Cumbrian County Councillors are being lobbied discreetly to refuse planning permission… It is thought that direct action might be counterproductive'.

The next issue of *Undercurrents* carried a full page report on the rally, but mentioned only FoE's participation and gave prominence to Walt Patterson's contribution: 'All in all FOE can be well pleased with the results of their first foray into direct action on the nuclear issue… FOE received a fair amount of publicity and they took the issue out of the realms of rarified abstraction and brought it to the people'. Again the *New Scientist* report of 29 April was not so supportive, talking about the bland character of the discussion, the minimal attendance by local people and the clear trade union support for nuclear power. However *Undercurrents* was concerned about the emphasis most of protestors put on safety issues, which 'in many ways is the issue on which nuclear power is least vulnerable'. It remarked that nuclear's safety record 'compared favourably with any comparable industry, as the BNFL spokesmen were quick to point out'. Once again *Undercurrents* felt the need to offer FoE advice, saying: 'FOE needs to broaden the issue to include economics, social policy and the whole question of the impact of technology on society, before they can hope to turn what is at the moment a minor protest into a major movement'.

BNFL's role in the FoE rally

However, unknown to anti-nuclear activists, the reason for FoE's hostility to 'direct action' was that FoE was being compromised by BNFL. According to Harold Bolter's recollections, he saw FoE's event as a public relations opportunity for BNFL and the chance to gain valuable publicity. He therefore 'decided to negotiate with Tom Burke, the Director of Friends of the Earth, and Walt

Patterson… to see if we could agree how their day of protest should be organized and managed'. Thus BNFL agreed to help FoE organize their visit to the site, both in its programme and facilities. BNFL agreed to provide speakers and equipment for a debate on the merits of reprocessing in order to keep the demonstrators occupied. As Bolter cynically admitted, 'if we refused to get involved in the event the demonstrators, finding themselves without anyone to argue with, would have time on their hands. They might then decide to try to draw attention to their cause in other, more mischievous ways. I had therefore agreed with Friends of the Earth that the interminable debate planned for the day of protest would take place on land owned by BNFL near the main entrance to Sellafield'.

FoE further agreed to BNFL's policing requests: 'The road in between this spot and the site's perimeter fence would be no man's land, which Tom Burke agreed his members would be told not to cross. Anyone who did could be treated as a potential troublemaker and arrested, with the support of Friends of the Earth'. BNFL then agreed to provide all the equipment, catering facilities, and toilets. Bolter wrote: 'I asked the Sellafield management to provide a platform, microphones and loudspeakers for the debate and tip off local agricultural show caterers to set up food and drink marquees, as I wanted to keep the demonstrators in well-fed good humour. We even provided portable lavatories and a crèche'.

FoE supporters were totally unaware of BNFL's role, just attributing the catering facilities to 'some enterprising locals selling coffee and hot dogs'. Finally Bolter had a leaflet prepared explaining reprocessing and 'arranged for the leaflet to be left on the train taking their supporters back to London'. Bolter was well pleased with his efforts, writing: 'The day went off better than we could have hoped'. He observed that one TV commentator, Martyn Lewis, did a piece for ITN news saying he was unsure whether he had been at a pro-nuclear or anti-nuclear rally. Finally Bolter commented that 'Friends of the Earth muttered something about making the Easter demonstration at Sellafield an annual event, but they have not held one since'. Overall he concluded 'it was a bit of a damp squib' for FoE. The ease with which FoE was compromised by BNFL is explained by Bolter because 'it was not in the interests of either Friends of the Earth or BNFL for there to be any violence. Both organizations wanted to come out of the event with their reputations for reasonableness intact… Their leaders coveted respectability for their views. Tom Burke and Walt Patterson believed that they could stop the reprocessing of overseas fuel, perhaps all reprocessing, by force of argument and assiduous political lobbying… and came close to doing so'.

The Windscale rally marked the beginning of FoE ascendancy over ConSoc in the anti-nuclear campaign, and also the start of the mass movement. From then on the anti-nuclear campaign was to be seen by the environmental and main-stream press (and future historians) as an FoE campaign.

Activists' criticism of FoE

London FoE's handling of this rally was criticized by some local branches, which typified the autocratic relationship between the centre and local branches over organization and tactics. Tom Barrance of Cardiff FoE in a letter in *Undercurrents* no. 17 wrote:

> There was little consultation with local groups over the decision to run a 'nuclear excursion' train... There was enthusiasm both among us and other FOE groups, for the proposed occupation of Torness power station site... however central FOE decided that there was too little press value in lying in front of bulldozers. Even direct action at the Windscale site was judged 'counterproductive' so all the press saw were 600 or so bored and cold demonstrators listening to a succession of uninspired speakers.

Barrance further complained about the high cost of the train fare, the lack of local representation apart from Half Life, and the small response from local people. He concluded: 'The rally appeared to fall seriously between the two stools of demo and debate. I would argue that the only way to get our point across without being outmanoeuved by BNFL etc. is to use direct action: demonstrations, pickets, occupations – somehow the public must be aroused and informed'. This call for direct action was consistently rejected by both FoE and ConSoc in favour of 'rational' debate and dialogue. Another letter in *Undercurrents* no. 19 from Ken Barker, again from Cardiff, welcomed Tom Barrance's comments and remarked on 'the deafening silence both before and after the event'. He complained about the lack of communication between FoE and other anti-nuclear groups, and the poor response from local people, and welcomed the possibility of an occupation at Torness. He commented on the lack of any anti-nuclear campaign writing: 'Following the Windscale Excursion it seems that there has been very little discussion on a national anti-nuclear campaign. The FOE energy workshop held at Whitsun [June 1976] hardly mentioned the subject... what is still lacking is sufficient consultation between anti-nuclear groups'.

He therefore proposed a meeting between active participants in FoE, ConSoc, Half Life, SERA, CND and other interested groups with the aim of forming 'a co-ordinated national campaign to actively prevent the nuclear industry's proposals going ahead'. This call perhaps resulted in the founding of the 'Campaign Against Nuclear Energy' based on Cardiff FoE, and the 'Alternatives to Nuclear Technology' based on Oxford FoE which had close links to the Oxford-based Political Ecology Research Group. All these three groups participated in NIN.

CANTO demo

Seven months after the Windscale rally came the first national demonstration. Organized by CANTO, it was held on Saturday 20 November 1976 in Trafalgar Square, London. It was well attended by celebrities but poorly attended by activists. Godfrey Boyle, the editor of *Undercurrents*, wrote in *The Windscale Controversy* (1983): 'The demonstration, though small in scale, succeeded in snatching a considerable amount of media attention mainly because of the fame of some of the participants: these included actress Diana Rigg, violinist Yehudi Menuhin, the Bishop of Kingston (Hugh Montefiore), television script writer Kit Pedler and Liberal Peer Lord Avebury'. These people were not only famous but also long-time activists. Lord Avebury was president of ConSoc from 1973-84, and a long-time critic of nuclear power, while Yehudi Menuhin was president of ConSoc from 1969-70. Kit Pedler was not only a TV script writer but also an anti-nuclear activist who wrote for *Undercurrents* and was to appear as a witness for the Windscale Appeal during the inquiry in 1977. Hugh Montefiore was active in the debate by the World Council of Churches on nuclear energy and was to chair public hearings into the fast breeder reactor in 1977.[77]

The poor attendance could have been because Czech Conroy, the FOE Energy co-ordinator, had publicly dissociated FoE from CANTO in a letter to *The Guardian* the day before the demo. This was because Hanson had announced in pre-rally publicity that the event was organized with the participation of FoE. This 'embarrassed and infuriated' FoE, particularly when the low turnout, of about 150 to 200 people instead of an expected 10,000, received nationwide TV coverage. FoE's mistrust of Hanson caused FoE to distance itself from the Windscale Appeal, of which CANTO was, in FoE's view, too dominant a member. This lack of interest and activity by activists is perhaps reflected in Undercurrents coverage of nuclear power during early 1977. At a time of extensive media coverage on the Windscale application, *Undercurrents* no. 20 in early 1977 had no articles on the nuclear debate, and none of the twenty letters published was on the anti-nuclear campaign. Again there were no letters on this topic in the next issue of *Undercurrents* (no. 21 in spring 1977).

The National Energy Conference

With no sign of the Windscale controversy dying down in the media, Tony Benn decided to hold another public debate on nuclear power, called the

'National Energy Conference', on 22 June 1976, again at Church House. ConSoc and FoE were invited to submit a joint paper, which was prepared by Walt Patterson and entitled 'Towards a National Energy Strategy'. Among the 470 members of the 'establishment' audience were the ConSoc president, Lord Avebury, and John Davoll, and Beryl Kemp (representing the National Society of Women). While ConSoc and FoE were on the inside there was a picket outside representing 'uninvited' anti-nuclear groups. These groups stressed that the promotion of alternative energy sources and energy conservation could mean many more jobs at a fractional cost of going ahead with reprocessing. Supporters of the picket included SERA, Half Life, the Ecology Party, the Nuclear Reactor Vigilantes, London Greenpeace, the Young Liberals, CND, several local ConSoc groups, Dr. Schumacher, and Sir Kelvin Spencer.[78]

While FoE concentrated on developing its 'insider' status and scientific credibility at 'elite' meetings, activists at ConSoc concentrated on building a wide ranging movement. As Beryl Kemp reported in the ConSoc 1976 *Annual Report*: 'Action is being taken to coordinate the efforts of many anti-nuclear groups in the UK and plans are afoot for demonstrations, silent protests, marches. Later it is hoped to combine with European activities and world-wide measures'. ConSoc participated in the formation of three umbrella groups, NIN (the Nuclear Information Network), the Windscale Appeal and Energy 2000. They also concentrated their efforts on their 'Safe Energy' petition, and on publishing letters in leading papers. The petition was modelled on the 'Clean Energy Petition' in the USA which had collected 400,000 signatures by July 1976. The aim of ConSoc's petition was 'to provide much needed evidence to the Energy Secretary that there is positive support for Safe Energy and opposition to Nuclear Power'. The petition was launched in early 1977, distributed widely even to Women's Institute branches, and had obtained 20,000 signatures by September 1977 with a target of 30,000 by November and 50,000 by the New Year.

ConSoc's proposed participation at the Windscale Inquiry and its anti-nuclear policies were opposed by some ConSoc members. Just before the inquiry opened, one letter in *Conservation News* no. 65 denied the existence of a pro-nuclear lobby in the UK and said: 'We follow the lead of FoE on their out and out opposition to nuclear energy which is motivated by their experience of the American scene'.

Windscale Appeal

The decision to hold a public inquiry was announced by Shore on 7 March 1977. Anti-nuclear groups however failed to agree on a unified presence at the inquiry under the banner of 'Windscale Appeal' (WA). As the ConSoc 1977 *Annual Report* commented:

The decision meant that the groups who had registered objections to the application had to decide on a plan of campaign and to this end an initial meeting was held at 9 Poland Street, attended by FoE, ConSoc, TCPA, CANTO, Civic Trust, Lawyers Ecology Group and others. It was not determined at this meeting whether legal representation was essential nor was there any certainty at that stage of sufficient money being available to fight the case. Agreement seemed to have been reached that the objectors should band together under the title of Windscale Appeal to avoid any one group's taking precedence. However, Friends of the Earth intended to confine their evidence to the necessity for and the economics of the reprocessing plant itself, and also wished to be solely responsible for organizing the case.

Nevertheless WA launched an appeal on 2 April for £30,000 to enable them to hire legal counsel and to defray other expenses involved in the expected three-month long inquiry. *The Guardian* urged, in an editorial of 15 April 1977, that the issues were 'so significant in their implications that the main opposition groups should be financed from public funds'. FoE estimated that it needed a total of £26,000: £15,000 for legal costs and accommodation, £2,000 for solicitor's costs, £3,000 for witnesses, £3,000 for research staff; and £3,000 for travel, telephone costs and other incidentals. Three weeks before the inquiry it had raised less than a quarter of this, and FoE was eventually to spend £100,000.

The announcement on 2 May by Sir James Goldsmith (brother of Edward Goldsmith, editor of *The Ecologist*) that he and his business colleagues would contribute £25,000 to fund opposition groups at the inquiry, created further controversy. It was not made clear to whom, when and how the money would be paid, nor whether this was a firm commitment or a fund matching exercise. The result was intensive competitive lobbying between groups, and its eventual distribution caused great bitterness. As the TCPA commented at the NIN post-mortem in November 1977, co-operation between groups was 'seriously damaged by the offer of funds and the attempt of various groups to get as great a share as possible'. FoE and the Windscale Appeal each got about £10,000, with £2,000 to the TCPA (who expected a great deal more), £2,000 to NNC and £1,000 to PERG.

The Windscale Appeal represented a wide variety of local and national groups. It consisted of CANTO, ConSoc, Cornwall Nuclear Alarm, the editorial group of *The Ecologist*, the Ecology Party, London Greenpeace, the Irish Conservation Society, the Society for Environmental Improvement and Wexford Nuclear Safety Committee. Most of these groups were part of NIN. Its committee, established 'in an informal way', was dominated by CANTO; out of its five members two were from CANTO (John Hanson and Edward Dawson), while Irene Coates represented both CANTO and ConSoc (as convener, land use and planning working party), with Beryl Kemp representing ConSoc and Julian Boles London Greenpeace. As explained earlier the presence of John Hanson. whom FoE distrusted and who was considered by Wynne as having an 'inability to co-operate' with other groups, meant that the WA was ostracized by other objectors.

ConSoc believed that the inquiry would be broad enough for it to be represented alongside other groups with disparate views, besides costing less than acting alone. It wrote in its 1977 *Annual Report* that it believed that the terms of the inquiry 'would be wide enough to allow the Conservation Society to put its own arguments for alternative energy sources and no expansion of nuclear power'. Irene Coates was appointed by the WA to co-ordinate the case and attend the inquiry. Her position was made impossible by internal conflict and as Wynne commented, 'there was no clear executive to prepare a case and brief its QC' with 'an approximately coherent case'. Her triple membership of CANTO, ConSoc and the WA undoubtedly led to conflicts of interest, and as the *Annual Report* stated: 'It should be made clear that Irene Coates acted in this [the inquiry] for the [WA] committee, rather than as a representative of the Conservation Society'. As the Annual Report remarked: 'Although the Conservation Society was by far the largest of the constituent bodies of the Windscale Appeal, we gained little publicity and our name was seldom mentioned'.

WA costs were about £20,000, of which only £250 came from ConSoc. David Widdicombe QC, assisted by Alan Alesbury, was engaged to represent them for a fee of £6,000, which he waived. WA presented evidence on the supposed need for nuclear power, its costs and risks, the disposal of nuclear wastes, the constraints imposed by the Euratom Treaty on the nuclear policies of the UK and on the greater use of district heating. Its witnesses in order of appearance were: John Davoll, Robert Blackith (Irish Conservation Society), Professor Ivan Tolstoy, Norman Jenkins, Colin Sweet, Irene Coates, Dr Barry Shorthouse, Professor Gordon Atherley, Dr Kit Pedler, Dr Charles Wakstein, and Edward Goldsmith.

The resulting fiasco

The WA's evidence and witnesses were not well received by either the inspector or the press, and ConSoc's reputation suffered compared to that of FoE, which rose. As the 1978 *Annual Report* commented, ConSoc 'had only limited influence in formulating the case and choosing witnesses. By the end of the Inquiry, press comment on the proceedings had dwindled considerably, but reaction to the Windscale Appeal's performance was not very enthusiastic'. Covering the inquiry for *New Scientist* Ian Breach wrote (20 October 1977) that WA's presentations had been disjointed and careless, though some of its witnesses had made able and credible submissions. Breach singled out the film presented by Charles Wakstein, one of the WA witnesses, as a particular example of its misjudgments that 'beggar explanation'.[79] FoE's case in contrast received praise from Breach, and the inquiry 'marked a turning point' for them. Reporting after the closing speech by its QC, Breach wrote (3 November 1977): 'No longer can FoE be regarded as an enterprising but rather unavailing conservationist lobby. After its case and this submission here, it will be seen as an important technical and political force'. His report was full of praise for FoE's conduct during the inquiry, from its astute appointment of Raymond Kidwell as its QC, and Oliver Thorold as his junior, to its selection of issues and witnesses. Breach, after examining who had the 'best' case against THORP, concluded: 'On an overall balance of form, content, style and timing, though, FoE emerges as the cardinal adversary in these hearings'.

The Guardian in its book *Windscale* (1978) on the inquiry ranked the protestors efforts. Some it said (implicitly referring to FoE) 'had presented their evidence expertly, with speed and precision', while others 'have failed miserably to have any impact on the hearing, showing themselves to be clownish and misinformed'. The Windscale Appeal, it remarked, 'should have been a powerful voice at the inquiry. In fact the group has been weak and blundering'. Beryl Kemp implicitly acknowledged this criticism when she wrote in ConSoc *Newsletter* no. 67 in late 1977: 'Some of the evidence presented by the objectors has, unfortunately, been more emotional and alarmist than factual and though it may have made the headlines it has probably not influenced the Inspector – indeed it has tended to irritate him and discredit the environmental case'.

The only favourable publicity for ConSoc was by Ian Breach, who was the former press officer of ConSoc. He commented favourably on submissions by John Davoll and Barry Shorthouse, and included Davoll's evidence in full in his book *Windscale Fallout*. The WA only raised £14,000 towards its total cost of nearly £20,000. About half came from Sir James Goldsmith, £2,000 from People for a Non-Nuclear Future, £1,000 from the Cheney Peace Settlement, but less than £3,000 from member groups of the WA. At the end of the inquiry it owed its solicitors nearly £5,000, and ConSoc (as part of the WA) was threatened with legal action for recovery of this debt. Eventually ConSoc settled the debt, but not without much internal recrimination between Irene Coates and council members and overall regret at the 'fiasco' of their involvement with the WA.

The aftermath

Roger Parker, the Windscale inspector, presented his report to Peter Shore on 26 January 1978. At first it appeared that the report would not be published before the Cabinet had reached a decision, despite Parker's wishes. This lack of public debate caused alarm. Ian Breach wrote in an editorial in the *New Scientist* on 26 January: 'Technically, he [Shore] is within his rights. Tactically, he is acting unwisely... the environmentalist lobby is morally justified in demanding that the Parker Report... be seen and considered by more than a handful of ministers, senior civil servants, and possibly some of those with a vested political interest in the application'.

There were appeals by MPs to Tony Benn, known for his views on open and responsive government, to exert his influence, and this together with lobbying from environmentalists ensured that the report was finally published on 6 March 1978. Parker recommended that outline planning permission for THORP should be granted to BNFL 'without delay'. Shore, introducing the report to the House of Commons on 6 March, found the report's conclusions 'persuasive and broadly acceptable' and said that he would normally have gone ahead and granted permission to BNFL. However he wished to give the House of Commons the opportunity to debate 'this unique set of issues' so a parliamentary debate was held on 22 March. After six hours of debate, the House voted by 186 votes to 56 in favour of THORP. There was another debate on 15 May and THORP was again approved by a decisive 224 votes to 80, though many speakers had qualms about approving it. In his book *Nuclear Entrapment*, William Walker commented twenty years after these speeches: 'Reading them today, one is struck both by their quality and by the air of unease that pervaded them. Two issues troubled the speakers above all others: whether THORP would encourage proliferation of nuclear weapons capabilities – and might damage relations with the US – and whether the decision that was being taken by Parliament would be irrevocable'.

Reaction to the Parker Report

The contents of the Parker Report took most of the participants in the inquiry by surprise. According to Michael Kenward – news editor of *New Scientist* – BNFL's opponents were staggered by the inspector's uncompromising acceptance of the case for expansion, while the company seemed equally amazed that its plans had come through almost unscathed. In an article of 9 March Kenward quoted Walt Patterson as saying, 'I didn't realize how fundamentally Parker was inclined to reject our line' as during the inquiry Patterson had the impression that Parker had understood the arguments on both sides. In the same issue Breach, reflecting Patterson's views, wrote that there would be dismay and even anger from those who were expecting 'a report that would, as near as realistically possible, reflect the detail and the character of their case... the report fails to do this and more seriously, misrepresents views and obfuscates or distorts the context in which those views were tended to the inquiry'. Breach singles out as evidence of misrepresentation the treatment of two principal witnesses, Brian Wynne and Peter Taylor, most of whose evidence is not cited. Another commentator, Jeremy Bugler, in *New Statesman* of 10 March, called it more a polemic than a report and said that seldom was there an 'inquiry report that cared so little for the appearance of even-handedness'. Parker, he wrote, by oversimplification of the issues, ended up 'by muddying the water', and it would not be surprising if the anti-nuclear movement now changed its approach.

FoE's reaction

The Parker Report was a bitter blow to FoE, who had high hopes for their case based on what they saw as their intellectual rigour and the perceived respect given to their case by Parker. As Robert Lamb, the FoE biographer remarked: 'The judge seemed to have understood their arguments and Patterson was optimistic about the chances of a win'. Patterson wrote much later in his history *Going Critical*: 'Those like Friends of the Earth who had come away from the inquiry itself feeling that they had made a strong case, and that Parker had taken it on board, could find in the ninety-nine terse pages of the Parker report no sign of their case whatever. Their arguments had not been refuted so much as simply ignored'.

Commenting in an interview nearly two decades after the inquiry, Patterson recollected that the 'Windscale report was the worst experience of my life, Parker kicked us in the balls'. Patterson however still believed that the inquiry would have been won by FoE, if other groups had not been involved and 'muddied the waters' with their evidence, allowing 'Parker to discredit our evidence'. As he concluded: 'I remain convinced that if it was just us versus BNFL at the inquiry, it would not have gone ahead. But other environment groups got on the (anti-nuclear) band-wagon and presented a total mish-mash of arguments. There was a smoke-screen of amorphous arguments, Parker could pick and choose, and tarred us with the same brush as the lunatic fringe'.

Patterson believed that Parker understood their evidence but when it came to writing his report he 'lost his brains' and that 'it was complete travesty of our case. It was a great shock when we lost; we were naive to believe our case would be considered on its merits. But we had to oppose it and to take part'.

Czech Conroy, the leader of the FoE campaign was stunned by the treatment of their case. As he confided to Robert Lamb: 'I thought the odds were stacked against us at the beginning... but we put together a very credible case. By the time that the inquiry ended, I reckoned our chances of victory at about fifty-fifty... What really stunned me was the incredible one-sided nature of the report. We had really believed that our ideas would be accurately represented and objectively considered – even if the final judgment went against us'.

As Tom Burke, director of FoE at that time, remembered in an interview: 'We believed we had a coherent case to put. We thought winning the argument would carry the day. This belief ended with the Parker Report. The Parker presentation of our case was a travesty, we were shafted. We felt a sense of betrayal and our confidence was shattered, it produced deep and lasting disillusionment'.

Mike Flood, another of the FoE staff involved with the anti-nuclear campaign from its early days, commented in an interview that the Parker Report had 'made me very angry, as our logical arguments were dismissed. We weren't expecting to win, but we had to take part as FoE had lobbied for an inquiry'. Flood believed that an important aspect of the inquiry was that 'we were gaining a platform for our views and our views were treated with respect, while their arguments were falling apart' and above all 'FoE was very interested in getting the facts straight'.

The Parker Report was not only a dismissal of FoE's evidence but also, more severely, called into question their strategy of protest based on expertise. As Patterson wrote some years later in *Going Critical*, the Parker Report 'shattered any illusions about the force of rational argument on nuclear issues... the report proved to be a numbing dismissal of every opposition argument, and could have been written without even holding the inquiry'. Writing three years later, in *New Scientist* (30 April 1981), Bugler accused FoE of being politically naive at Windscale, saying FoE failed 'to realise that it was going before a hostile bench'. As Bugler recollected: 'I still have fresh in my mind the winded astonishment of the members of the Friends when they heard the news of the Parker Report. They were like men who had been asked to speak at a meeting for a Worthy Cause and found that while they were doing so, someone had picked their pockets'.

He contrasted their optimism with 'more worldly-wise environmentalists' who had 'rumbled Parker even while he was taking evidence'. Bugler put the failure of FoE down to its belief in working within the political system, saying 'this naivety is accompanied by a commitment to politicking, ...accompanied by a march-them-up-the-hill belief in the political system and the trustworthiness of Whitehall'. Robin Grove-White, who was active in the anti-nuclear campaign but in his terms only 'very peripheral' to the Windscale Inquiry, believes that the inquiry was a turning point for FoE. Nearly two decades after the event he judged, in an interview, that 'FoE had run a brilliant campaign and they were devastated by the result. They thought they had won the arguments. I knew FoE hadn't done as well as they thought they had... maybe I was wiser in the ways of the world'. Grove-White was more experienced, and perhaps more cynical, about public inquiries, and thus was not surprised by the result. His longer experience, dating back to 1971 to the campaign against the proposed Shell oil terminal in Anglesey and from his work with CPRE, meant that he better understood the planning system. As he remarked in 1996 on the shock of the inquiry result and its impact on FoE: 'I was prepared for it but this was the first inquiry for them. Several witnesses wrote to *The Times* saying their evidence was travestied in the Parker Report but I couldn't take it seriously, I just thought this man is a bigot'. He concluded on FoE naiveté, saying 'FoE was flummoxed, they were innocent, they were full of terrific energy and fun, and felt it was a bit of an adventure. It was quite an adjustment afterwards'.

In April 1978 FoE published a bitter critique called *The Parker Inquiry*, detailing the inadequacy of the official report, which was presented to Shore the day before their demonstration. They accused Parker of being biased in favour of the nuclear industry's plans, and of being selective in his use of evidence. While the inquiry had set 'a landmark in nuclear policy making', its 'hasty and erratic judgments' had produced a 'polarisation rather than a moderation of the debate'. Thus FoE concluded the report had made it more difficult to see how public confidence in the worth of participation in such lengthy and costly procedures could easily be restored. Or in Patterson's simple words, 'The Parker Report polarized the nuclear issue in the UK essentially beyond any hope of recovery'.

ConSoc's reaction

ConSoc was not so disillusioned, not having such high hopes about their evidence, and seeing the Windscale Inquiry as no more than a 'rehearsal' for the long-awaited inquiry into the first commercial FBR. As Beryl Kemp remarked in *Conservation News* 69: 'We lick our wounds and prepare for the Fast Reactor Inquiry' and 'It makes depressing reading for environmentalists because the Inspector has failed completely to understand the real issues at stake'. Furthermore Parker was 'sceptical and dismissive of any optimistic claims by the objectors of the ability of conservation and alternative energy systems to meet the energy gap should it arise'.

However the Parker Report won approval from some members of ConSoc, led by John Fremlin, for its lack of bias. In a letter to *Conservation News* in the summer of 1978 he defended the impartiality and expertise of the two assessors at the inquiry, saying: 'It is the bias of the ignorant that worries me. Well-meaning ignorance is defenceless when faced with propaganda'. In the following issue (no. 70) Fremlin compared the efforts of the nuclear lobby to those of the coal or oil lobby. Commenting on Ralph Nader's actions in stopping nuclear power stations in the USA, which had led to more fossil fuel being sold, he wrote: 'I don't suppose for a moment that Ralph Nader was himself paid by the oil companies or that anyone, or any organisation, in Britain has been paid by them. Why should they pay for propaganda when a whole lot of well-meaning people will do it for them for free?'

There was a chorus of criticism in much of the mainstream press about the Parker Report, with many witnesses feeling that their evidence had been totally ignored. However despite its widely perceived faults, the failure of objectors to get their views across to Parker was to some extent self-inflicted because of the often poor quality of their evidence. The *Guardian* book (1978) commented that 'the objectors will probably all agree they could have performed better if they had kept their attack short, sharp, free from obscure meanderings and, most important of all, not been so numerous'. Ian Breach, in a *New Scientist* editorial (10 November 1977) after the end of the inquiry, argued that while many of the submissions were powerful and credibly articulated, their cases were 'damaged by the rambling, shambling and repetitive format in which they found themselves'. He believed that the environmentalists had failed their first major test at Windscale, and the future of the nuclear debate was bleak. In his book on the inquiry, *Windscale Fallout*, he wrote that 'the politics of nuclear power and reprocessing will increasingly be those of disillusion, anger, frustration and mistrust. We shall all be the losers'. However Brian Wynne was more positive about the outcome of the inquiry, saying in *New Scientist* (3 August 1978) that 'despite the unambiguous repudiation handed out by Parker, objector groups enjoy greater credibility and influence than ever before'.

Overall the inquiry established FoE's leadership of the anti-nuclear movement with the media. As Martin Ince, a freelance journalist, remarked in *Undercurrents* no. 28 'One thing that Parker's inquiry proves is that nuclear opposition in Britain is Friends of the Earth'.

FoE demonstration

The Parker Report dealt a severe blow to FoE's pride and anti-nuclear strategy based on lobbying and debate. In early 1978 they decided on a change in tactics and called for a mass demonstration in London for 19 March which was postponed until Saturday 29 April, following a two-month ban on marches in London. The march was well attended, and was by far the largest anti-nuclear power demonstration seen in Britain. According to the accounts in *Undercurrents* and *Vole* (no. 9) it was the new, more militant, voices speaking out in support of non-violent civil disobedience that drew the crowd's approval, not the message of continued conventional protest from Tom Burke of FoE and John Davoll of ConSoc. *Undercurrents* no. 28 reported:

The mood of the demonstrators who packed Trafalgar Square was probably best captured by Arthur Scargill, the Yorkshire Miners President, who told the crowd of more than ten thousand that: 'If it needs civil disobedience to stop nuclear power, then we shall have to have civil disobedience' – words which were greeted by the loudest cheers of the afternoon (though the Master of Ceremonies for the day, FoE Director Tom Burke, was seen to shudder visibly as the words 'civil disobedience' were uttered.

The crowd was urged to contemplate picketing and occupation of nuclear facilities if the Windscale expansion was sanctioned by parliament and existing plans for nuclear-powered generation were continued. Calls for direct action worried some commentators, with Michael Kenward in a pre-demonstration editorial in *New Scientist* (4 May) warning against the 'Eco fascism' of anti-nuclear extremists who 'wanted to go to any length to bring a halt to nuclear power in Britain'.

Call for direct action

The call for a more radical approach than that adopted by FoE or ConSoc had been made for some time by local activists and by the movement press, such as *Peace News* and *SCRAM Energy Bulletin*. It was further reinforced by FoE's failure at Windscale and adverse comment about groups working 'within the system'. In an article in the *New Statesman* (10 March 1978) Jeremy Bugler commented that 'until now the anti-nuclear forces in Britain have shown themselves willing to protest "within the system". In sharp contrast to French or German opponents, they have argued that society can be persuaded from nuclear power. But Parker does not offer dialogue... it will not be surprising if the anti-nuclear movement here now changes its approach'. Similarly in a review of future anti-nuclear tactics, Wynne remarked (*New Scientist*, 3 August 1978): 'Disillusioned with the rough treatment meted out by Parker, FoE and others are beset by considerable internal conflicts as to whether the streets are not a better political forum than the committee room'. Furthermore he commented: 'FoE's members have been highly critical of their leadership's moderate line and its attention to lasting credibility with the establishment'. There were fears that participation in the widely expected Fast Breeder inquiry would lead to a similar defeat.

Dave Elliott, active in SERA and in mobilizing trade unions for the anti-nuclear campaign, agreed writing (*Undercurrents* no. 28): 'The case for more effective, direct action through civil disobedience has never been stronger'. Assessing the pros and cons of participating in the expected Fast Breeder inquiry, he wrote that participating in an inquiry did have tactical advantages, in terms of gaining publicity for counter arguments. However it seemed likely that 'just as at Windscale wider arguments will be listened to, but sidestepped' and that 'no amount of technical argumentation seems likely to influence the official decision makers'. He thus proposed an intermediate strategy, combining limited participation in inquiries with 'careful grass roots organising in the community as a whole'. He believed 'direct action' would not stop the construction of reactors, but their value was that they were mainly 'public relations' activities, aimed at 'demonstrating the strength and commitment of the movement, and thereby putting pressure on the decision makers'.

In response to Elliott's article David Pearce, who was conducting the Windscale Assessment and Review Project, made (*Undercurrents* no. 31) 'a brief plea for giving the establishment one more chance' with the proposed Fast Breeder inquiry, due to take place in 1979. He argued that an inquiry was an important means of soliciting information, 'and that much can be done with this information which may be surprising in terms of the results it yields'. Pearce's worry was that if this inquiry was boycotted an opportunity for generating valuable information through a 'quasi-adversarial process' would be missed, and the 'don't knows' in the public would not be informed.

In a major review of anti-nuclear tactics post-Windscale (*New Scientist* of 27 July 1978), Ian Breach commented that: 'the existing environmentalist movement faces its most difficult period, in Britain at least, over the coming debate on fast-breeder development. If they lose their case against the FBR, their chance of widening and popularising their concerns will dwindle. Knowing this the government *could* pull the rug from beneath their feet'. Breach urged the centre-left environmentalists to build a much broader based movement, including the 'unemployed, the deskilled, the homeless, the aged poor, racial minorities and others' through working with such groups as Age Concern, the Lucas Aerospace Combine Shop Stewards' Committee, the Claimants' Union and the Conservation Society. Elliott too was in favour of a broad-based movement, writing in *Undercurrents* no. 28: 'The task now is to start to build a broad, but politically aware, anti-nuclear movement, which steers between co-option and manipulation by meaningless government inquiries on the one hand, and over zealous, adventuristic, direct action on the other'.

While Breach and Elliott, and many other activists, pondered how to build a broad-based anti-nuclear movement, Wynne was wondering (*New Scientist* of 3 August) about the opposite: 'It may be that movements with political interests "extraneous" to the nuclear debate will now exploit the gap between leadership and members to raise wider issues and to recruit anti-nuclear protestors to their politics'.

This 'broad-based' strategy was to be put to the test over the next year at the proposed AGR site at Torness in Scotland. Here a peaceful occupation of the site in early May 1978 by several thousand people was followed by direct action by activists from the Torness Alliance. They attempted to obstruct building work and from October began a month-long occupation of Half Moon cottage, which was scheduled for demolition. These actions received much publicity and caused argument about non-violent protest in *Peace News*.

The election of the Conservative government of Margaret Thatcher in May1979 – which promoted both nuclear power and nuclear weapons – further enhanced the appeal of direct action to activists disillusioned with the moderate approach of FoE. As an impassioned editorial in *New Scientist* (29 November) – most probably written by Ian Breach – said, 'Environmentalists who appeal to their version of good sense, to rationality, to good husbandry on planet earth, and so on are falling on deaf ears: those with access to the levers of power are not always swayed by rational debate'. The editorial therefore welcomed the creation of the Anti-Nuclear Campaign (ANC) 'that could become more influential and more decisive in British politics than any other since the suffragettes'. However this (anti-FoE) view was challenged by Michael Kenward in an opposing *New Scientist* editorial (20 December) which warned of the rejection of sensible criticism in favour of 'guerilla' activity, saying that there was 'plenty of room for technically informed critics' and that the anti-nuclear movement 'may not be winning the war but it is winning some battles'.

The main beneficiaries of this more militant mood were not the existing anti-nuclear movement, nor established environmental groups, but Greenpeace. This new organization was from the late 1970s able to win vast publicity and thousands of members by its uncompromising emotive appeal to popular sentiment, combined with non-violent guerilla tactics. As Harold Bolter commented on the anti-nuclear opposition post-Windscale Inquiry: 'Friends of the Earth's anti-nuclear campaigners lost heart. Many of their natural supporters began to look for something more telling than the reasoned arguments against nuclear power and reprocessing provided by Tom Burke and Walt Patterson. They were ready for action – and Greenpeace stepped in and gave them what they wanted. Argument was quickly replaced by anger and aggression'.

New organizations arose to harness this 'anger and aggression' and further diminished the appeal of the measured FoE approach. The ANC and other groups were to protest from 1979 against both nuclear weapons and nuclear reactors, and thus directly link the peace and the environment movements. Activists from the Torness Alliance transferred their 'direct action' skills and feminist ideology of the associations between nuclear power and patriarchy to the women's peace camps at Greenham Common in the early 1980s. This overt linking by feminists of patriarchy and nuclear power as a systemic source of violence and oppression also played upon a long-standing divide in the public mind over the merits of technological progress.

In the USA the linking of the peace and the environment movements was aided by the foundation of Mobilization for Survival in September 1978, whose efforts were greatly assisted by the vast media coverage of the accident at the nuclear power station at Three Mile Island (near Harrisburg, Pennsylvania) in March 1979. A spokesman for the connection of nuclear power and nuclear weapons was anti-nuclear activist Michio Kaku, who wrote in the US magazine *The Guardian* in May 1979:

> Nuclear power and nuclear weapons are two sides of the same coin. They are controlled by the same people, produced by the same corporations and serve the same political and financial interests. They give off the same radioactive poisons, generate the same deadly wastes that nobody yet knows what to do with. And both threaten catastrophic destruction. The people who brought us Hiroshima now bring us Harrisburg.

The stage was now set for the long and bitter battles against nuclear power and weapons in the 1980s – something outside the scope of this book. The next chapter turns its attention to what happen at THORP when it finally got constructed. A prize British asset or a white elephant?

Chapter 18

THORP constructed

After the House of Commons endorsed Parker's report in early 1978, BNFL proceeded swiftly to sign contracts with Japanese and Swiss utilities in August 1978, and then with German utilities in 1980 and 1983. THORP however only began construction in 1985 and was not completed until 1992. Before it could start operating, BNFL needed various regulatory authorizations for its discharges of radioactivity which it expected to be rubber stamped by the relevant ministers without significant scrutiny.

However the energy world had changed greatly since THORP was approved. It had been built mainly to serve a (perceived) foreign demand for reprocessing but now (in 1992) its main customers such as Germany were trying to walk away from their contracts. Its strategic rationale in the 1970s – the provision of plutonium to fuel the FBR – had collapsed as the FBR was no longer considered a viable reactor. Since there was no demand for plutonium, reprocessing was now unnecessary, and its production only aggravated the problems of nuclear waste disposal and nuclear proliferation. Instead of THORP being risk-less 'easy money' for Britain, it now appeared a huge liability and this concerned the Treasury who became increasingly worried in the early 1990s about THORP's financial risks. Now, with BNFL's requests for authorizations, there was a chance for the government to put this 'white elephant' out of its misery, and stop a project that had lost all validity.

But such a step was far too painful for the Conservative government and a civil service 'entrapped' in long-standing commitments. Despite their deep internal misgivings, they decided to approve BNFL's requests. As William Walker, a nuclear historian argued in his book *Nuclear Entrapment*: 'The best and most obvious explanation is that the commitments to THORP were too deeply embedded, and the commitments to sustain the commitments too determined, for

the case for THORP to be overthrown through protest and weight of argument…
the ship of state moved itself inexorably towards its destination'.

The government's only problems were how to justify the decision, in case it
was challenged in the courts, and how to have public consultation without
holding a fresh public inquiry. They therefore carefully crafted a 'legitimation'
strategy to demonstrate that operating THORP would be politically and legally
sound. Officials were instructed to ensure that the decision to approve was robust,
brought some net benefit to the UK, and would not embarrass ministers. The
government's behaviour was driven largely by its vulnerability to judicial review
if THORP were not submitted to another public inquiry. The government needed
to present the judge with evidence of public consultation, but the debate had to
be so managed that the 'public' could gain no influence on the final decision. The
endgame of this charade of public consultation came on 28 June 1993 with a
parliamentary debate and a vote of 147-43 against a Liberal party motion
opposing approval, with much of Labour abstaining. This debate was one of the
low points in THORP's history. The Liberal motion was too long and complicated, and it was ambushed by the Labour MPs Dale Campbell-Savours (the MP
for Workington, neighbouring Sellafield) and Tam Dalyell. The debate showed
how effective BNFL and its lobbyists were in mobilizing parliamentary support.
Walker remarked that whilst its PR was 'often clumsy and ineffectual, the
company ran a carefully orchestrated programme since late 1992 to curry
support among parliamentarians'.

After this parliamentary approval, John Gummer announced, on 15 December
1993, his official approval for BNFL's application, saying that there was 'sufficient
balance of advantage in favour of the operation of THORP'. Greenpeace imme-
diately applied for a judicial review, which opened on 7 February 1994. The
verdict was delivered on 4 March by Mr. Justice Potts who found in favour of the
ministers on all relevant points. So in late March 1994, BNFL received the authorizations it had applied for two years previously, and the radioactive
commissioning of THORP began.

Ineffectual opposition

The opposition in the 1990s to THORP was similar to that in the mid-1970s
– loud but ineffectual. This time it was led by Greenpeace, rather than FoE, again
supported by local groups such as CORE (Cumbrians Opposed to a Radioactive
Environment). The national and international branches of Greenpeace took all
possible steps available within the law to injure THORP's prospects between
summer 1992 and March 1994. However its efforts had little effect, as it could
not use one of its most effective tactics, namely consumer boycott (which it was

to do with great effect with Shell over Brent Spa in 1995). Also THORP never became a party political issue, with parliament showing little interest during this period. Indeed the Conservative government was always confident of the support of the Labour party – THORP had been approved by the last Labour government. Most important of all, the MP for the Copeland constituency that embraces Sellafield was Dr Jack Cunningham, a prominent member of the shadow cabinet and the party's national executive – he helped anchor the Labour party to the government's position. Then there was strong trade union support for THORP, as the Sellafield site was heavily unionized and BNFL had long been skilled at winning influence in the Labour party through the trade unions. Finally THORP was too local, arcane and complex an issue to sway whole electorates. The Conservative and Labour parties sensed that THORP carried few electoral risks, despite huge media coverage which played up fears of cancer and childhood leukaemia.

Thus in 1994 the British government decided, after long and agonized internal debate, to allow the plant to operate. As William Walker remarked: 'In taking the decision, it knew that a Rubicon was being crossed. As THORP's start-up would contaminate the plant with radioactive materials, society could not just walk away from THORP if the decision came to be regretted'.

Walker lays blame on opponents for not coming up with a coherent alternative to BNFL's proposals for reprocessing spent fuel. Their emphasis – as with the 1977 Windscale Inquiry – was on killing THORP, not on devising more acceptable alternatives to the handling of foreign spent fuels delivered under contracts. Greenpeace and the other protest groups were unable to forge a consensus on the alternatives, with Greenpeace (Germany) wanting a more hard-line position than Greenpeace (UK), so the safe option of evading the issue was chosen. As a result the so-called 'win-win option' (where THORP was mothballed, which had been discussed in the DoE) did not achieve any prominence in the public debate.[80] The failure of both Greenpeace and the government to give serious scrutiny to this option meant, according to Walker, that 'they neglected to explore and enunciate an exit strategy and they did not test the plausibility of their preferred outcome'. Although Greenpeace was unsuccessful in stopping THORP in 1994 it is likely that no judicial review would have occurred without Greenpeace's initiative, the threat of which concentrated the mind of the government greatly. Thus without the efforts of environmental groups, the public and private debates about THORP would have been stunted, and the decision itself would have been less carefully considered.

THORP's performance

THORP was conceived in the 1970s, constructed in the 1980s at a cost of £1.8 billion, and operates to provide plutonium fuels for foreign and domestic nuclear energy programmes that are long since dead and buried. Since its opening THORP has performed fitfully. In the first five years of its operation from 1994 to 1999 it operated at about half planned capacity, and reprocessed some 1900 tonnes of spent fuel, or 28% of the planned throughput. It was closed for most of the year from spring 1998 to spring 1999, and there have been technical problems with excessive tritium discharges. Reprocessing has therefore had a slow and erratic beginning. The future looks no better, for BNFL told *The Guardian* (26 August 2003) that it was planning to close THORP around 2010 – when it has fulfilled its current contracts.

Thus market economics may achieve what its operators, its customers and numerous governments seem incapable to doing – laying to rest one of the largest and most controversial industrial facilities ever conceived in Britain. The end result of reprocessing has not been commercial cornucopia, but the accumulation (so far) of 75 tonnes of plutonium and 3,336 tonnes of uranium, all stored and closely guarded but with no obvious use.[81] While environmentalists may have failed to stop THORP, they have hopefully laid the grounds for better decision-making and public participation in nuclear power issues. Eventually their efforts succeed, in that 'white elephants' do finally get put out their misery as those (in government and the civil service) committed to their survival pass away.

Chapter 19

Why anti-nuke?

W̶hat was it about nuclear power that stirred up such great passions in the 1970s after it had been generally acknowledged in the 1960s by conservationists as better for the environment than coal mining or dam building. Why the loss of support for an industry that claimed to be the shining hope of a brave new technological world, that was supported by all shades of political opinion from radical trade unionists to conservative businessmen, that was the ultimate in the 'progressive' dream?

This chapter pulls together the arguments and analyses made in this book, in particular arguing that existing accounts have been inadequate in some respects. It makes the following general points about the anti-nuclear movement:

- There were very mixed motives for the opposition to nuclear power, not all 'environmental'.
- There were two distinct yet related national campaigns against nuclear power in Britain, the first in the mid-1960s, the second in the mid-1970s.
- The origins of opposition to nuclear power are complex, but are best summarized as 'nuclear fear'.
- There has been a tendency in the histories to concentrate on the role of FoE, and ignore other anti-nuclear groups and activists.
- The role of women activists has been neglected, as a result of an emphasis on technical and policy debate.

Finally it poses the question of whether the current promotion of solar (and renewable) energy has the characteristics of 'energy utopianism', and if so, is this likely to be successful at creating a sustainable future? Or will the opposition to its prescriptions (such as wind farms) emerge as it did to nuclear utopianism?

Anxieties over radiation

The numbers involved in the anti-nuclear power protest were small in the USA and most European countries until the early 1970s, but they did have an impact on public opinion despite strong support for nuclear from government, business, the trade unions, most professional organizations and political parties. That the small anti-nuclear power movement could shift public opinion was a source of puzzlement to early nuclear commentators. For they believed that nuclear energy enjoyed high public support until the early 1970s, and that support fell away due to a shift in values by many people towards less material goals. This view was challenged by Ian Welsh, in his book *Mobilizing Modernity*. He argued against the view that there was 'a past golden age of public acceptance or at least public quiescence' and that 'the public acceptance of nuclear power only became problematic in the early 1970s'. Instead there has been since the birth of atomic energy a profound, but concealed, ambivalence towards nuclear technology amongst the public. Ambivalence and public opposition is based not on recent value changes but on long-standing social, cultural and moral attributes.

The public expressed its anxieties over nuclear power through explicit questioning of reactor safety and the dangers of radioactivity, despite dismissals of such fears by inquiry inspectors. Moreover the refusal of the public authorities, such as the CEGB, to acknowledge these fears made it impossible for them to shake off their reputation as unresponsive and aloof organizations. Furthermore their inability to provide a convincing answer to the frequently asked question (dating back at least to the first nuclear inquiry at Bradwell in 1956) of why, if nuclear power stations were so safe, were they built in remote areas? And why were there such detailed emergency measures in the event of any release of radioactivity? During the campaign at Stourport letter writers raised the issue of nuclear safety – one mentioned the Windscale accident of 1957. This theme of fear about nuclear safety dominated the campaign, despite reassurances from experts engaged by the county council. Cllr. Eric Higgs, the secretary of the APS – the anti-nuclear power station committee – was quoted as saying (*KT&N* 5 June 1970): 'We are objecting because we do not want this in Stourport. There will always be an element of fear, and we don't want to live with it'. Another objector, Cllr. Betty Gazard, a prime mover behind the campaign, expressed this fear of radiation when she said: 'None of us know how much we shall receive, or if we shall receive any, or if it will do us any harm'.

So what were the roots of these anxieties and fears?

Origins of anti-nuclear power attitudes

The reasons for the emergence of the anti-nuclear power movement in the UK in the 1970s are obviously complex. Like all technological assessments made by the public the reasons were only partly based on what scientists would call a 'rational' evaluation of the risks and benefits. There were of course many 'technical' concerns, but the movement was at heart an emotional response to 'nuclear fear', to dystopian images of an atomic future laid down since the beginning of the nineteenth century. That the movement emerged in the early 1970s was in part due to opportunistic reasons – the public collapse in confidence in government institutions and authority, as well as internal dissent within the nuclear establishment.

The nuclear proponents in the 1970s persisted (as with the 1957 WHO study) in seeing opposition to nuclear power as simply 'emotional'. Len Brookes, a leading propagandist for the UKAEA, wrote (*Atom* April 1975) that anti-nuclear feeling was 'a largely irrational reaction stemming from deep emotional conviction rather than any dispassionate analysis of the problems and the practical options for dealing with them… it still relies for its wider support on attitudes that are largely emotional and irrational'. In common with elements of Labour party opinion – as with Anthony Crosland – he saw the anti-nuclear movement as predominately elitist and middle-class. This opinion continued through to 1980 with Frank Chapple, the leader of the electricians union, denouncing 'hysterical voices of environmentalists, ecologists and sundry political opportunists, who exploit public ignorance'. In an after-dinner speech he said (*Nuclear Energy* vol. 19 no. 2):

> They rewrite all our known experiences with nuclear energy, embellishing every detail, exaggerating every mishap and behind this smokescreen (sometimes euphemistically described as a low energy strategy) they skillfully conceal the fact that the logical outcome of their policies will, at worst, leave us with a shortage of energy around the year 2000 and, at best, lead us first to stagnation and then to reduced standard of living.

The source and dynamic of the anti-nuclear opposition may have been emotional but they were not irrational. The author, I believe, who comes closest to identifying the causes of opposition is Spencer Weart. In his book *Nuclear Fear: A History of Images* he identifies four main themes that have influenced the way people thought about nuclear power. These were:

1.   The technical realities of reactors, both the economic opportunities and the hazards, as seen by scientists and transmitted to the public. From these realities particular 'facts', such as the hypothetical effects of low-

level radiation in the event of an accident, were selected and stressed.

2. The social and political associations of nuclear energy, especially ideas involving modern civilization and authority. These associations explain what happened when reactors became a condensed symbol for all modern industrial society.

Weart argued that nuclear power was singled out for this symbolic role largely as a result of:

3. The old myths about pollution, cosmic secrets, mad scientists and apocalypse that were historically associated with atomic power and radiation – indestructible myths with deep psychological resonances.

4. The threat of nuclear war, never for a moment forgotten.

Thus to Schumacher the old myths of scientists trespassing on forbidden territory were still valid. In an article 'Economics in a Buddhist Country' written in 1955 he said that atomic energy was 'a prospect even more appalling than the Atomic or Hydrogen bomb. For here unregenerate man is entering a territory which, to all those who have eyes to see, bears the warning sign "Keep Out"'. For John Davoll (CN 47) it was mistrust of scientific experts and pessimism about the ability of humanity to manage vast amounts of nuclear wastes safely for millennia. 'Can any society so dependent on the advice of experts... have any hope of arriving at informed and democratic decisions on the most vital questions that face it?'

Reasons for opposition

What exactly were the campaigners against nuclear power opposed to? Their motives were diverse, ranging from NIMBY concerns through to opposition to capitalism. Motives for opposition to nuclear power can be divided into four categories:

1. NIMBY – local opposition to any large-scale development nearby.

2. Vested interest – opposition by the coal miners fearful of their jobs.

3. Intellectual – based on aesthetic, ecological, ethical and economic reasons.

4. Opportunistic – an opportunity by political groups to attack government policies.

NIMBY concern

NIMBY opposition is easy to understand and identify, and forms the bedrock of local opposition to proposals for nuclear reactors. In its early days it was termed the 'amenity issues' lobby – people opposed to developments in unspoilt countryside. Partly this was out of aesthetic concern by urban intellectuals for preservation of outstanding scenery, and partly due to the rural aristocracy and middle class attempts to prevent economic developments that might undermine their privileges. Hence Anthony Crosland's allegations in 1970 about the conservation lobby being hostile to growth and having a manifest class bias. For all large-scale developments in rural areas will bring benefits and costs to different sections of society, and nuclear power was no exception. Communities were often divided over the issue. The more vocal and well organized middle classes, organized into ad hoc amenity societies, were however better able to put their views across and drown out less well articulated working class support. This occurred in 1956 over the Bradwell and Hunterston nuclear stations.

In the late 1970s, with the mushrooming of local groups opposed to proposed nuclear power stations, there were again accusations of NIMBYism. Gavin Weightman, a sympathetic commentator, wrote (*New Society* November 1979) that 'the people who join such groups would probably oppose a coal-fired power station or an airport on the same site, and many are not perhaps, strictly anti-nuke'. Local middle class opposition (in the early 1980s) to nuclear power stations was again accused of NIMBYism, portrayed by the media as 'piece-meal, irrational responses based in parochial concerns'. However such local groups were defended against these charges if they were willing to oppose nuclear power nationally and become part of a network of anti-nuclear organizations, such as the Anti-Nuclear Campaign or SCRAM. Thus while local nuclear opposition can be seen as NIMBY, it can redeem itself (in the eyes of sympathetic commentators) if it is willing to become part of a national network opposing nuclear power. To cynics this is simply 'greenwash' – the adoption of an environmental position to further self-interest.

Vested interest

The coal miners had a clear economic or vested interest in opposing nuclear power, due to their determination to stop job losses in their industry. The opposition of the coal miners, their union, their supporters in the Labour party and the National Coal Board (NCB) to the expansion of nuclear power at the cost of coal in the mid-1960s has been overlooked by most historians of the anti-nuclear movement. The campaign led by Alf Robens and Fritz Schumacher ended in defeat in 1968. Robens, the NCB chairman, wished to protect the coal industry from a too rapid shut down, and argued vociferously that nuclear power was being unfairly subsidized. Particularly significant was the role played by Fritz Schumacher, then chief economist at the NCB, and later after the publication of *Small is Beautiful* (1973) a leading spokesman for the environmental movement. Schumacher had a long-standing opposition to nuclear power dating back to the mid-1950s and he was actively involved in the 1970s campaign. Opposition to nuclear power continued in the 1970s from sections of the NUM with Arthur Scargill, its president, involved in the creation of Energy 2000, which was part of the opposition at the Windscale Inquiry in 1977.

Intellectual dissent

There was criticism of nuclear policies and siting based on aesthetic, ecological, ethical and economic reasons but until the mid-1970s these were largely confined to academic journals and small movement publications. There had always been critics of the government's nuclear policies, but these were brushed aside by the nuclear institutions determined to proceed with their plans. Dissent initially came from industry insiders, concerned about the feasibility of too rapid an expansion of the industry, followed by criticism from academics critical of nuclear economics and the AGR reactor choice. There was little public discussion of radiation or safety issues, until Schumacher raised the issue in a speech in 1967, which proved highly controversial but was quickly forgotten. Few people saw the linkages between nuclear weapons and nuclear power, and whilst the public were fully aware of, and campaigned against, the dangers from radiation fallout from nuclear testing, they appeared unaware of radiation emissions from nuclear reactors.

This changed in the early 1970s due to the books from the United States – Curtis and Hogan's *Perils of the Peaceful Atom* (1969) and Gofman and Tamplin's *Poisoned Power* (1971). These publicized the safety hazards and dangers of radiation emissions from nuclear reactors. American campaigners quickly adopted these arguments against nuclear reactors, after initially relying on ecological ones about 'thermal pollution', disturbance of habitat and aesthetic damage to the landscape.

The concerns of Sternglass over the hazards of low-level radiation were given wide publicity by the national press in Britain, and were raised during the debate over the proposed Stourport reactor in 1970 by some of the consultants to the county council. Very rapidly scientific dissent over 'safe' limits for radiation exposure, and also the unresolved problems over radioactive waste storage, became translated by eco-activists into the prime reason for the public to oppose nuclear power.

Opportunistic dissent

Intellectual critics of nuclear power were also concerned with issues of democracy and equality, seeing nuclear institutions as examples of remote and overbearing bureaucracies that threatened civil liberties and must therefore be curbed. This criticism of the 'nuclear state' built on previous criticism of the modern technocratic state and the power of corporations, such as by Theodore Roszak and Lewis Mumford. Nuclear power in the USA thus became a rallying point for student radicals and other social critics, particularly after the end of the 'campus wars' and demonstrations against the Vietnam War in the early 1970s. Previously the new (and the old) left had shown little interest in environmental or nuclear power issues, sometimes attacking it as a distraction from more serious social issues. In the USA protest against nuclear power (like environmentalism) was seen as a middle-class provincial movement concerned with NIMBY issues, whereas in Europe it was seen as a means of uniting peasant farmers with students. American nuclear commentators Irvine Bupp and Jean-Claude Derraine in their book *Light Water* (1978) commented on this difference:

> American opposition to nuclear power crosses traditional ideological and party lines. To a certain extent the same appears to be true in Europe. But there is an important difference. In Europe, opposition to nuclear power has become a means to achieving more fundamental ends by true radicals – who often label themselves 'ecologists' – desiring basic change in the economic and social fabric of their societies. In other words, opposition to nuclear power is one of the most effective rallying points around which European social critics gather.

This can be seen in Britain by the gradual adoption of the anti-nuclear power cause by the new left, starting with the Socialist Workers Party in 1976. It also attracted religious, peace and women's groups who had a long-established concern with nuclear weapons.

The US anti-PWR campaign

FoE's anti-nuclear campaign of the 1970s had its origins in the US anti-nuclear movement, since FoE started in California and both its main anti-nuclear campaigners, Patterson and Lovins, were (north) American. Their criticism of nuclear power, particularly the PWR, rested heavily on US evidence and expertise they had gained from attending US hearings. Furthermore it can be argued that this American influence is symptomatic of the general American dominance of youth culture in Britain since the late 1960s, which manifested itself in terms of music, fashion and politics. The motivations and goals of US and British critics were not always similar. In the USA critics need not be 'environmentalists' and vice versa. Bupp and Irvine made these interesting comments on nuclear critics:

The motivation and the real objectives of many American and European nuclear critics are not completely clear, but a few generalizations seem safe. A difficult thing for Europeans to understand about American nuclear critics is that some of the most prominent and influential are neither political radicals or even 'ecologists' as that label is commonly used in Europe. For them opposition to nuclear power does not go beyond rejection of a specific technology.

For instance Duncan Burn, in *The Political Economy of Nuclear Energy*, was a scathing critic of the UK programme and the British AGR, and was very much in favour of the American LWR (on economic grounds). In contrast Amory Lovins and Walt Patterson, in their early criticism, were against the LWR on safety grounds but in favour of the AGR. Patterson argued (*Environment* December 1972) that the AGR was not only 'more efficient', but also 'more environmentally sound and considerably less suspect on the score of safety'. FoE's submission to the Science select committee in 1973, together with media appearances and press articles by Patterson and Lovins, were seen as attacks on the safety record of nuclear power (in the form of the PWR). This appalled nuclear supporters who until then had counted on conservationists for support in their campaign against the poor environmental and safety record of coal. In his review of the British anti-nuclear campaigners, R. F. Pocock spoke of this bias: 'It is perhaps significant that the environmentalists groups who were openly critical of the postulated hazards of the nuclear industry were far less outspoken about the actual deaths of coal miners or about the proven long-term dangers of coal wastes such as engulfed Aberfan in 1966 and threatened Trehafod ten years later'.

However British concerns over the dangers of coal mining were of no concern to American disputes over PWR safety, which can be seen as a civil war waged on British soil by largely American experts from a branch of an American organization, FoE. A war ultimately directed against the attempt by the Americans to make the PWR the dominant nuclear reactor worldwide. Thus it is important to remember in the words of Bupp and Irvine that 'nuclear critics are not always environmentalists, nor are all environmentalists opposed to nuclear power'.

The two anti-nuclear campaigns

In the decade from the mid-1960s to the mid-1970s there were two distinct yet related national campaigns against nuclear power in Britain. The first in the mid-1960s was the campaign by the NCB and the miners against the replacement of coal by nuclear power. It was strictly a vested interest campaign, with a clear agenda which used conventional political methods of lobbying. The second was the less focused campaign against nuclear power in the mid-1970s organized by a wide variety of environmental and political groups, using new tactics of street protest. Despite their very different agendas and styles they used very similar arguments against nuclear power. These arguments (detailed in the many policy histories of nuclear power) were:

1. Lack of transparency in nuclear costings.
2. Lack of accountability of the nuclear executive.
3. Excessive secrecy in nuclear discussions.
4. Misuse of national resources.
5. Opposition to the electrification of the economy.
6. Dangers of excessive centralization and monopoly power.
7. Existence of desirable alternatives.

New arguments added in the 1970s were:

8. Safety concerns: hazards of radiation and potential for accidents.
9. Problems of radioactive waste disposal.
10. Dangers from nuclear proliferation and terrorism.
11. Threats to civil liberties from the nuclear state.

The first two of these new arguments had been raised by Schumacher in his 1967 lecture, based on concerns raised in the US literature, but at that time they were considered outrageous slurs on the competence of the British nuclear industry. The last two were 1970s concerns brought about by the use of terrorism by various political groups and some countries in the late 1960s and early 1970s.

Both campaigns failed to achieve their goals. They made little impact on the political support given to nuclear power. The 1960s campaign was politically far stronger. It was led by Lord Robens, a former Labour minister and chairman of the largest nationalized industry, and was backed by many Labour MPs and the NUM, the most powerful trade union. It had unequalled access to the highest levels of government and the civil service. It could, and did, put its message across to those in positions of power. It was inside but not listened to. The prevailing opinion was that coal was finished, nuclear was the future. The only important issue was how to run the coal industry down as humanely as possible. This was the view of most MPs, the media and even most of the NCB. The anti-nuclear campaign was seen as a purely vested interest. It also failed to attract any support outside the coal industry, all sections of society being convinced that nuclear was the technology of progress.

The 1970s campaign had far less political support and access. But it had more media coverage and some public support. It had style but no power. It influenced public opinion somewhat but failed to attract established organizations, such as business, trade unions or the professions. It was outside and ignored by the establishment. Summing up the achievements of environmentalists, Jeremy Bugler wrote:

I am driven to the conclusion that it has been a decade of considerable intellectual achievement and success in consciousness-raising, but one of great practical failure... Environmentalists in the UK have failed to prevent Windscale's growth... or indeed an expanded nuclear programme; they have nobly opposed motorways and delayed them for a few years, but in the end the Civil Service has come back to build them.

This power of the civil service and politicians, to ignore and rebuff attacks on their policy, was also remarked on by Alf Robens who wrote in his autobiography *Ten Year Stint* (1972): 'It saddens me to think that the country has suffered because the advice I and my colleagues (notably Schumacher) gave was ignored by the civil servants and the politicians. We were regarded as bloody nuisances at the time'.

Schumacher continued to be a 'nuisance' to policy makers, taking an active part in the 1970s campaign. His involvement in both campaigns makes him almost unique amongst conservationists who in the 1960s were almost wholly in favour of nuclear power, viewing it as less polluting than coal. The reason nuclear power was to fall out of favour with them is almost certainly to do with the voicing of public concern about the dangers of radiation, and this was evident at Stourport in 1970.

Stourport protest

The local debate over the proposed Stourport nuclear power station in 1970 contained all the elements that were to re-occur in the mid-1970s – local risk, national benefit, radioactive dangers and amenity concerns. Despite the local councillors not being able to find any credible scientific or legal grounds on which to oppose the station, they nevertheless persisted in their opposition fuelled by a fear of the risks involved (from possible release of radiation). It could be maintained that this was a NIMBY excuse, that the opposition were looking for any grounds, the more alarmist the better, on which to oppose the building of a power station. Opposition could be ascribed purely to an 'anti-modernization' fear, mentioned by a local paper as existing in the nearby town of Bewdley over a by-pass proposal. Undoubtedly there was such an element, and the proposed nuclear station would have had a massive dominance over the town of Stourport. However there was an existing coal-fired station at the site, and the Stourport council had lobbied the CEGB for several years to build a new station there. However, even councillors who opposed the nuclear station were not against the building of a new coal-fired plant. Cllr. Eric Higgs, secretary of the APS, said (*KT&NS* 5 June 1970): 'A power station will come, but do we want it to be nuclear or conventional? We would rather have a conventional one'.

Thus it was opposition to a nuclear station, rather than to any power station, that motivated the local councillors. And the driving force behind that opposition to nuclear was fear of radiation, of an accident that would contaminate them. Higgs summed it up as not wanting to live with an 'element of fear'. It is impossible to determine why there existed this fear of radiation amongst the sober, respectable and supposedly level-headed citizens and councillors of this small town in rural Worcestershire. Perhaps it was long-standing, dating back to the controversy and public debate about the consequences of nuclear fallout from atmospheric tests in the 1950s and early 1960s. For radiation, after the use of the atomic bomb, had never had a good image. It was something to be feared, a 'deadly invisible menace', or as academic Tim O'Riordan commented in the journal *Catalyst* no. 12 in 1986: 'Radioactivity falls into a specialised class of substances that strikes fear into the public mind because it is unknown, unfamiliar and is associated with lingering death which the innocent cannot avoid'.

Perhaps this image lay dormant in the minds of the councillors for many years, until an event came for it to be confronted. This, combined with dislike of the CEGB's patronizing and arrogant ways, and of nuclear hype and phoney claims by its supporters, laid the basis for what was in essence an emotional, rather than a rational, opposition. It was based on pessimism about the risks of nuclear power, and typical of this attitude was the objection by Kidderminster Borough Council on the grounds of the possibility of radioactive fallout in the event of accident. They commented 'that the risk is too great a price to pay for benefits'. This fear was despite all existing scientific opinion in Britain about the desirability and safety of nuclear power. The consultants to the county council, Leslie and Shaw, remarked: 'We feel that, provided the Council is satisfied about the radiation safety aspects of the proposal, a nuclear station is much to be preferred to one fired by a conventional fuel such as coal'.

The public fears raised at Stourport about the dangers of radiation were carried over into the public inquiry into the proposed reactor at nearby Portskewett a year later. Thus the debates at Stourport and Portskewett in 1970 to 1971 showed that an increasing minority were not satisfied by experts about the 'radiation safety aspects', and this concern was to become over the next five years the principal rallying point of the anti-nuclear movement (but not of its movement intellectuals).

It is highly unlikely that Schumacher's ethical critique of nuclear power made in 1967 was known amongst the general population, as his views were quickly dismissed by the media. It was the anti-nuclear criticism coming from the United States that found a much wider audience – books by John Gofman and Arthur

Tamplin – and undoubtedly the British press publicity given to Ernest Sternglass. By 1970 media coverage of environmental issues, a willingness by some people to support amenity over economic growth, and the residual effect of fears of radioactivity stemming from concern over nuclear fallout in the early 1960s meant that proposals for nuclear power stations got a more critical reception than previously.

Ironically the site at Heysham, where the CEGB did succeed in building an AGR (without even an inquiry) in 1969, was to be the birthplace of the 1970s anti-nuclear movement. Here was to be the formation in 1975 of the first organization, Half Life, explicitly devoted to oppose nuclear power.

Chapter 20

FoE hagiology

Most of the history of the 1970s anti-nuclear campaign has been written from an FoE perspective, chiefly by Walt Patterson. Another major source of information is *Promising The Earth* – Robert Lamb's hagiology on FoE. This draws heavily on material provided by Patterson – his previous accounts in *Nuclear Power* and *Going Critical* and interviews – with a few quotations from Czech Conroy, in its seven page account. In discussion of the composition of the anti-nuclear movement, FoE is almost invariably the only organization mentioned. Lamb does not mention the campaigns by ConSoc, Half Life, NIN or any of the other groups while Patterson only mentions them in his first account in 1979 and not in later publications. Their accounts are exclusively FoE-centric. Czech Conroy in his history of the Windscale campaign makes no mention of ConSoc's involvement (and hardly any of Half Life), stating only the FoE branch's involvement. While Roger Williams in his history mentions Half Life's campaign three times it is FoE that gets by far the most extensive coverage. Similarly Crispin Aubrey in his account *Meltdown* mentions Half Life twice and FoE extensively (and, to be fair, many anti-nuclear groups in the 1980s as his book is mainly on the campaign against the proposed PWR at Hinkley).

Thus the existing histories of the anti-nuclear campaign are one-sided. While they detail the complexity of the organizations and people on the pro-nuclear side, on the anti-nuclear side there is just FoE, led by Walt Patterson with help from Amory Lovins and a few others, able to call upon an amorphous mass of 'environmentalists' to demonstrate. The history thus becomes David versus Goliath, a handful of people battling against the multi-headed nuclear hydra. This myth is reinforced by Patterson's aversion to the contribution of others less intellectually gifted than himself, who he claimed 'muddied the waters' with their evidence, allowing Parker to discredit FoE's evidence. To him the efforts of the other anti-nuclear groups were a big embarrassment, and responsible for the failure of FoE to achieve success at the Windscale Inquiry. That the failure of FoE was, in the words of their contemporary critics, because they were politically naive and too keen on becoming part of the establishment has never been addressed by them.

The dominance of FoE in the accounts written by Walt Patterson, particularly his 1985 book, together with the media attention given to FoE as the premier environmental group opposing nuclear power (achieved by its articulate performance at the Windscale Inquiry) has meant that anti-nuclear histories have been little more than FoE hagiology. It has been my intention to rectify this, by showing that the anti-nuclear movement in the 1970s was a great deal more complex and diverse than the written accounts indicate.

FoE expertise v ConSoc democracy

The Windscale Inquiry made FoE but broke ConSoc. It is somewhat ironic that the methods FoE used in its success – elite lobbying together with rational and highly technical argument – were the antithesis of the values it supposedly stood for as an organization – decentralized, informal and inclusive. In contrast ConSoc stood by its organizational values: democratic decision-making – with members voting and a committee structure – and espousal of environmental values.

FoE was never a democratic organization. It had supporters not members. It was run by the head office, with the director having total control over policy and staff, particular under Tom Burke. Head office ran campaigns, not branches. This was in contrast to ConSoc where members could put forward a policy resolution at the annual general meeting and then if it was accepted could start campaigning. Thus it is no surprise that the anti-nuclear movement started at ConSoc, as it was better able to reflect the concern of members. There may have been similar concern in FoE branches but they had no mechanism for articulating or acting within FoE until head office decided on an anti-nuclear campaign (in late 1975).

FoE head office was an avowedly elitist organization, more concerned with maintaining their reputation for technical competence (and the access that went with it) than combining forces with other groups. This is typified by their decision before the Windscale Inquiry not to ally themselves with any of the other groups but to present their evidence alone. Thus Brian Wynne in his account of the Windscale Inquiry, in *Rationality and Ritual*, characterized much of the conflict amongst groups as due to 'FoE's unwillingness to co-operate with other groups except on its own terms, and a resultant feeling that the career of FoE as a semi-establishment group took precedence over the more radical sentiments of many of its members'.

As Walt Patterson remarked to Robert Lamb about the inquiry:

> If it had been left to us, there was no way that the anti-case could be ignored. I think that it was a successful tactical move by BNFL to invite everybody into the room so that they could get under each other's feet and clutter up the arguments. What happened, there was a huge smear of arguments ranging from the very well worked out to the wildly angry and exaggerated, which left room for Parker to do what he did, to tar everybody with the same brush and say it was totally over the top, indefensible and so on, invoking everybody at the same time.

The FoE approach won the organization favourable press publicity and enhanced the staff's reputation amongst fellow professionals but it did not win it friends amongst the other fellow organizations (of NIN) and was inimical to building a mass movement. Perhaps this suspicion, even envy, did not matter to FoE. It was at least admired and acknowledged (by the press) as the leader of the anti-nuclear campaign. ConSoc in contrast decided to become part of a broad umbrella of groups at the inquiry. Its decision to delegate handling of its wide ranging evidence to the Windscale Appeal committee was a disaster for ConSoc. It won no favourable publicity and lost prestige to FoE. But as Wynne put it: 'The objectors were in an impossible position: if they took the wider view, the inquiry would be swamped and Parker would not take them seriously; if they took the narrower approach, they would disappear from the political debate.'

This conflict between expertise (or rationality) and democracy (or emotion) always plagues campaigns and organizations. To be successful a balance needs to be struck. Or according to Martin Ince (*Undercurrents* no. 28), a balance between the 'rather ill-matched types of activist, in a word, the public inquiry type and the street campaigner'. Without the first, found at head office, there is no credible case and no favourable publicity; without the second there is no mass organization, campaign or movement. The anti-nuclear power movement of the 1970s used FoE's head office as their brains and a widely disparate group of environmental and radical activists as their feet. Its success was that it produced in the three years, from March 1975 to April 1978, a movement capable of mobilizing over ten thousand people to march in London, and several thousand to camp the weekend at Torness. Its tragedy, or irony, was that this movement had nowhere directly to go – its focus after the expected Fast Breeder inquiry was scrapped. As new issues arose, the movement merged itself from 1979 into such 'direct action' organizations as the Torness Alliance, CND, ANC and Greenpeace. Both ConSoc and FoE were unable to capitalize on this new mood of militancy because their leaders, John Davoll and Tom Burke, were personally opposed to direct action.[82] The result was that ConSoc went into a long period of decline as members drifted away, and it was wound up in 1987. FoE however continued to expand and was to play a significant role at future nuclear inquiries at Sizewell and Hinkley. Like the Windscale Inquiry, these two were similarly lost.

FoE retrospective

The Windscale Inquiry has assumed that status of an epic battle, nearly won, amongst FoE hagiographers. As Czech Conroy reminisces about Windscale to Robert Lamb: 'It's true that we lost, but we gained a great deal in the process. There was a massive raising of public awareness about nuclear issues because of the daily media coverage of the inquiry and there was a big jump in the credibility of Friends of the Earth. They couldn't dismiss us as ecofreaks any more'. Conroy also points out that it established a significant precedent, that of holding a public inquiry when considering any major nuclear development and this delayed the progress of the nuclear industry, particularly over plans to build the Sizewell PWR. Furthermore Conroy believes that FoE's anti-nuclear campaign 'was one of the most effective campaigns that Friends of the Earth ever ran, especially when you consider what we were up against. Britain was a world leader in nuclear technology, with both civil and military programmes. To stop the expansion of the civil industry in only twenty years of campaigning was an incredible feat'.

Walt Patterson, commenting on the Windscale Inquiry two decades afterwards in an interview, believed that if FoE had not opposed THORP and gone to the inquiry, BNFL would have started building in 1976, with the plant perhaps opening in 1982. That would have resulted in the accumulation of far more plutonium, and there also would be reprocessing plants in other countries. He believed that FoE's success was that 'we delayed it by 15 years. We delayed the programme. We achieved a lot and feel proud'. Robert Lamb was similarly upbeat about their achievements: 'In retrospect, FoE's presence at the inquiry had done enormous good, raising the organisation's profile to a level it had never previously achieved and marking it out as the credible voice on nuclear issues. It had severely undermined the case for nuclear power, rattled a powerful but complacent industry and challenged the conventional wisdom that nuclear waste should be reprocessed'.

However Joe Weston in his account *The F.O.E. Experience* believed that the impact on the FoE movement was devastating. 'The Windscale issue had taken up a whole year and had resulted in many other campaign areas being reduced in terms of their profile and the amount of resources spent on them. The campaign had cost the organization far more than the £100,000 in lawyers and expert witness fees used during the course of the inquiry. For all that time, effort and expenditure the organization had, apparently, gained nothing. The groups had been asked to raise money for the campaign and at the end of the day, had received nothing in return'. Weston remarked that the FoE head office believed that having lost, the campaign was over. Many activists in local groups however did not accept defeat so easily and continued to oppose local reactors. The energy campaigners of FoE felt excluded from these informal local alliances, such as the Torness Alliance, and saw them as a threat, 'as if FoE was in danger [of] losing its lead over the anti-nuclear movement'.

Brian Wynne doubts that the inquiry changed anything in nuclear policy. Some environmental campaigners did however benefit, learning about doing research, putting evidence forward, forming international links, and creating a name for themselves. As Joe Weston perhaps cynically remarked, inquiries do 'provide campaigners with a stage on which they can display their talents and their arguments'. The inquiry did not raise the profile of nuclear issues amongst the public as there was little media publicity, especially in the mass circulation press, because inquiries are arcane and intimidating to most people. Wynne in an interview two decades later, said he had no illusions as he 'never expected to win but he didn't expect it to be fixed. There was enough evidence for the inspector to write an anti-nuclear case but Parker had a predisposition to accept evidence from government bodies as they were considered accountable and democratic'.

However FoE in its desire to put a compromise case, and establish credibility with the government, was outflanked by Parker at the inquiry. The inquiry was widely seen by both the press and parliament as a battle for or against nuclear power, but because FoE refused to join the other groups in mounting an anti-nuclear case, and instead argued for delay and for technical alternatives to reprocessing, the end result was that although they were 'the most seasoned and erudite of the objectors' their policies were not seen as a credible political option. While Jeremy Bugler in his review of FoE's first ten years (*New Scientist* 30 April 1981) gave them praise for their high quality work and advocacy on nuclear issues, he believed their Windscale campaign was a failure due to their political naiveté, remarking: 'If nuclear power has suffered rebuffs, it has been almost entirely from its own mistakes, not through public opposition'.

A retrospective view by William Walker was more positive. Although FoE 'suffered a significant defeat' at the inquiry, the government's victory was at a heavy cost to the future development of nuclear power in Britain. He concluded that 'THORP became an important symbol of anti-nuclear discontent in Britain' and 'the manner of the decision fed public mistrust of nuclear power which helped impede its further development in the UK, a cost that did not figure in the government's or industry's calculations'.

Emotion versus expertise

Media praise for FoE at the Windscale Inquiry was based on their use of a rational approach. A plea for the acceptance of 'emotion' as a valid premise for opposition to nuclear power was made however by Peter Taylor of PERG, in his evidence (reported in Breach's *Windscale Fallout*). Taylor argued that we lived in a society ruled by rationalists and reductionists because of our dependence on advanced technology, and therefore there is seldom an opportunity to discuss fundamental alternatives:

> If one were to say 'I oppose nuclear power because it threatens the very stuff of life', one would be branded as emotional and incapable of rational argument... there is far more scope for interaction if one says 'I oppose nuclear power because the disposal of long-lived actinides to geological formations of unproven permeability or long-term structural stability cannot guarantee against radiological hazards due to concentrations in the food chains.'
>
> Thus, anyone who can actually say that sentence and get the words right (they do not necessarily have to understand them) may become a front-line protagonist. The others must stay at the back and maintain a dutiful silence whilst their 'champions' do battle.

Taylor illustrated this point by the testimony to the inquiry by Barbara Fish, from a small religious commune, who posed simple questions such as 'Why must we always live with the blunderings of science all in the name of progress?' and 'How dare we "mere mortals" reason that our industries and refrigerators are important enough to put the balance [of nature] at stake?' As Breach commented: 'Such non-scientifically assembled questions lack credibility in a controversy dominated by economic and political "fact"'. He further noted that groups like FoE, for all their use of specialist witnesses and technical evidence, draw for their support on a very large number of people like Ms Fish and her fellow commu- nards. These people feel, instinctively, that 'things' are going wrong and that 'the move towards dependency on nuclear power is potentially the final mistake'.

To the mass of activists the inquiry was an obscure ritual of expertise in which their 'champions' took part. It was not part of their political campaign. As Wynne remarked: 'It is easy to forget that the Inquiry parties were a very small élite minority of activists for or against some aspect of nuclear energy... The "envi- ronmentalist" may have been in there fighting, but he too was an alien expert more akin to the industry's experts than the ordinary person'. Brian Wynne writing (*New Scientist* 28 June 1979) after the Windscale Inquiry remarked that participation in public inquiries for less experienced and less organized groups and individuals becomes more daunting the more technical they become. 'Representation becomes less and less meaningful the more inquiries are highly elite, professionalised and semi-private exchanges between strongly committed adversary parties. More elaborate inquiries may buy more natural justice for professional objectors only at the expense of even more alienation of the wider public'.

Green Alliance

The tendency to focus on the views of the 'champions' – the expert and the elite – rather than the ordinary activists was further reinforced by the founding of the Green Alliance. This was founded by Liberal peer Tim Beaumont at a meeting on 23 December 1978 at the National Liberal Club, attended by thirty people. John Davoll and Irene Coates from ConSoc attended, but Davoll did not join. By March 1979 it had 85 members, mostly the 'great and the good' from the environment groups, with hardly any overlap with the NIN activists. An exception was Irene Coates who initially refused to join because of 'the gross disparity of sexual representation' as only two of the initial thirty people invited were women. By March 1979 there were more women, 9 out of 85 members, a sex ratio of only 10%, far worse than ConSoc in the 1960s. There were women activists – 10 out of the 41 member groups of NIN had a woman as contact point. There were 8 women in the 29 people attending the 'Windscale post-mortem' NIN meeting in November 1977, but they were overlooked.

This neglect of women activists is now being rectified, particularly in the USA, by environmental historians. They are starting to examine the role of early women activists, in such books as Vera Norwood's *Made from This Earth* (1993) and Mary Breton's *Women Pioneers for the Environment* (1998). These detail campaigns against pollution from energy sources – ranging from the nineteenth-century campaign by the Ladies Health Protective Association of Pittsburgh against coal smoke, to Mary Sinclair's long and ultimately successful campaign against the building of a nuclear power station in Midland, Michigan in the early 1970s. However nothing similar exists as yet for women activists in Britain.

Chapter 21

Soft energy utopianism

A s we have seen in Chapter 5, energy technologies have been associated with a mixture of utopian and dystopian viewpoints. Utopian views initially dominate society, but over time there is a wavering of support as promoting institutions fail to deliver the expected benefits and the disbenefits become apparent. Sometimes there is a sudden shift in support for a technology, as sections of society abandon it in response to the impact of new ideologies. This can be seen with nuclear power in the early 1970s, where conservationists and many elements of the new left quickly switched their allegiances from nuclear to solar energy. Often driven by dystopian fears of a 'nuclear state' they joined the campaign by the new environmentalists against the expansion of nuclear power.

Solar energy, soon to be rechristened renewable energy, was to be the way forward to a 'sustainable' world. This energy utopianism started in the early 1970s with the AT movement, and its latest expression is in books such as Hermann Scheer's *The Solar Economy* (2002). In it, he claims that the adoption of solar energy will cause a revolution that will replace corporate ownership of fossil fuels with local ownership of renewable energy sources leading to a more democratic and egalitarian society.

So is solar energy to be the future energy source? Ultimately does it matter? Does choice of energy technology really affect how society is organized? Is Scheer correct when he claims that history shows that new technologies, including electric grids and nuclear power, have caused a technological revolution that has reshaped the world? The history of energy utopianism in fact, I believe, shows the opposite. New energy sources do not automatically bring desirable social and economic change. Rather it is the way society is organized, or in the classic marxist phrase 'who controls the means of production', which determines the scale and nature of the impact of energy technologies. All energy utopianism suffers from the flaw of 'technological determinism' – a point made by social critics of technology in the 1960s.

In an era of global capitalism energy sources are likely to be part of a globally-traded energy network. This could be the 1960s vision of Buckminster Fuller for a world-wide electricity grid based on solar energy coming from deserts, or more recent ones for a hydrogen economy based on wind, hydro or nuclear power. Whilst economic control of energy systems may still rest in corporate hands the physical infrastructure of energy generation is however becoming more decentralized, under the impact of much smaller-sized generating plants. For the UK this trend will undoubtedly continue, strengthened by government policy in its White Paper *Our Energy Future* (2003) for far more renewables and decentralized electricity generation. Whether this energy decentralism will bring the social and political benefits desired by solar supporters is uncertain, and is certainly not automatic. Unless there are other government policies in support of political and economic decentralization, the world solar enthusiasts aspire to may remain solely in the realm of utopian or science fiction – or occasional 'islands' of greenness always vulnerable to conquest by outside forces.[83]

A nuclear revival?

So does the history of the anti-nuclear power movement in the 1970s have any lessons for nuclear protest in the twenty-first century, or teach us anything about why people protest against new energy technologies? Does the current protest against wind farms have anything in common with previous protest against nuclear power stations? Can the promoters of wind learn anything from the experiences of the 1970s nuclear promoters? Or does the civil service always get its way in the end, despite a few years of delay caused by opposition?

The wind power debate is similar in rhetoric and tone to the nuclear power debate in the 1970s, although public opinion is more in favour of wind power than it ever was of nuclear power. Just as the nuclear lobby in the 1970s could not conceive of or accept the fact that people did not want any nuclear power stations (and wanted to close down existing ones), so much of the wind lobby is unable to accept the rationality of opposition. For instance the BWEA (British Wind Energy Association) uses the same language as was used by nuclear promoters in the 1970s. They complain about 'misconceptions and misinformations distributed by groups aiming to stifle wind energy development completely' and that 'a small but vociferous number of people have generated a disproportionate amount of press coverage'. Similarly wind promoters call for education, well-informed debate and local consultation and an end to NIMBYism. The technology is different but the debate and language is the same.

Just as the nuclear establishment was split in the 1970s, so there are divisions amongst 'environmentalists' over wind farms. This division could reflect NIMBY concerns, with some local and regional groups in areas where wind farms are planned being opposed, whilst all the big national groups are in favour. Long established regional groups, such as the CPRW (Council for the Protection of Rural Wales), are seeking the protection of the countryside based on the aesthetics of landscape, whilst national groups (RSPB, WWF, FoE, Greenpeace, SERA etc.) are united in seeing wind power as a symbol of a sustainable Britain. What needs to be understood is that opposition to wind, like that to nuclear, is partly a political protest against the (social) vision of its promoters. Some people simply do not like or want a 'sustainable Britain' or a 'soft energy future', just as some did not want a 'nuclear state' or 'breeder economy' in the 1970s. This protest can be considered as partly intellectual and partly opportunistic, and is supported by those sections of the media opposed to the energy and social policies of the current Labour government. Wind, like nuclear, power is promoted not just as an energy technology but as a means by which its supporters can achieve their social goals, hence the utopianism of Hermann Scheer.

So could anti-wind campaigners shift the emphasis of environmentalism towards the old preservation ideas (or perhaps deep green approach) of the protection of 'nature' rather than 'Man'? Could conservation groups concerned with rural protection, such as the CPRW and the CPRE, return to their 1960s support for nuclear power with their argument that it is better to damage already blighted urban areas rather than 'unspoilt' rural areas, with its unspoken assumption that it is better to subject people to some hazard rather than sacrifice nature? [84] This seems unlikely, for much as they may wish for an energy source that does not 'spoil' the countryside, the options are limited. Nuclear is too expensive and is only lukewarmly supported by government, whilst fossil fuels are accused of causing global warming. A nuclear revival is unlikely, at least for the UK, unless:

- Renewables fail to deliver technically and economically, causing the 'environmental establishment' to become split on energy policy.
- The public become disillusioned with the promises of solar utopianism.
- Minority countryside concerns are taken up by the urban majority.

The UK nuclear industry will however not quickly fade away. There will be a continuing role for places such as Sellafield in nuclear decommissioning and waste disposal. These places will be a mausoleum for future generations of the twentieth century's nuclear dream. Likewise the dream of a utopian society will remain with us, but the energy source through which it is expressed will change over time. The old vision of nuclear power may be over, but the dream of a utopian energy technology still lives on. Renewables are the current favourite, but whether they will be any more successful than nuclear power remains to be seen.

Bibliography

Adams, W. M. 1996. *Future Nature: A vision for conservation*. London: Earthscan.

Allaby, Michael 1971. *The Eco Activists: Youth fights for a human environment*. London: Charles Knight & Co.

Aubrey, Crispin 1991. *Meltdown: The collapse of the nuclear dream*. London: Collins & Brown.

Basalla, George 1982. 'Some Persistent Energy Myths'. In *Energy and Transport: Historical Perspectives on Policy Issues*, eds. George Daniels and Mark Rose. London: Sage.

Bolter, Harold 1996. *Inside Sellafield*. London: Quartet Books.

Boyer, Paul 1985. *By the Bomb's Early Light: American thought and culture at the dawn of the atomic age*. New York: Pantheon Press.

Boyle, Godfrey and Peter Harper (eds.) 1976. *Radical Technology*. London: Wildwood House.

Bramwell, Anna 1989. *Ecology in the 20th Century: A history*. New Haven: Yale University Press.

Braunstein, Peter and Michael Doyle (eds.) 2002. *Imagine Nation: The American counterculture of the 1960s and 1970s*. London: Routledge

Breach, Ian 1978. *Windscale Fallout: A primer for the age of nuclear controversy*. Harmondsworth: Penguin.

Bunyard, Peter 1981. *Nuclear Britain*. London: New English Library.

Burn, Duncan 1967. *The Political Economy of Nuclear Energy*. London: Institute of Economic Affairs.

Calder, Ritchie1962. *Living with the Atom*. Chicago: University of Chicago Press.

Carey, John (ed.) 1999. *The Faber Book of Utopias*. London: Faber & Faber.

Clarke, I. F. 1979. *The Pattern of Expectation 1644-2001*. London: Jonathan Cape.

Clute, John and Peter, Nicholls (eds.) 1999. *The Encyclopedia of Science Fiction*. London: Orbit.

Coates, Peter 1998. *Nature: Western attitudes since ancient times*. Berkeley, Ca.: University of California Press.

Cohn, Steven 1997. *Too Cheap to Meter: an economic and philosophical analysis of the nuclear dream*. Albany, NY: State University of New York Press.

Conford, Philip 2001. *The Origins of the Organic Movement*. Edinburgh: Floris

Books.

Elliott, Dave 2003. *Energy, Society and Environment*. London: Routledge.

Eyerman, Ron and Andrew Jamison 1990. *Social Movements: A cognitive approach*. Cambridge: Polity Press.

Ford, Daniel 1982. *The Cult of the Atom*. New York: Simon & Shuster.

Fountain, Nigel 1988. *Underground: The London alternative press 1966-74*. London: Routledge

Gould, Peter 1988. *Early Green Politics: Back to nature, back to the land and socialism in Britain*. Brighton: Harvester Press.

Hall, Tony 1986. *Nuclear Politics: The history of nuclear power in Britain*. Harmondsworth: Penguin.

Harrison, Harry 1980. 'Inventing New Worlds'. In *Future Imperfect: Science fact and science fiction*, Rex Malik (ed.). London: Frances Pinter.

Hay, Peter 2002. *A Companion to Environmental Thought*. Edinburgh: Edinburgh University Press.

Hays, Samuel 2000. *A History of Environmental Politics since 1945*. Pittsburgh: University of Pittsburgh Press.

Herring, Horace 1978. *Alternative Technology Directory*. Brighton: Pebble Press.

Hilgartner, Stephen, Richard Bell and Rory O' Connor 1982. *Nukespeak: Nuclear language, visions and mindset*. San Francisco: Sierra Club Books.

Holdgate, Martin 1999. *The Green Web: A union for world conservation*. London: Earthscan.

Kimber R. and J. Richardson 1974. *Campaigning For The Environment*. London: Routledge & Keegan Paul.

Kirk, Andrew 2002. ' *"Machines of Loving Grace"*: Appropriate Technology, Environment and Counterculture'. In *Imagine Nation: The American counterculture of the 1960s and 1970s*, eds. Peter Braunstein and Michael Doyle. London: Routledge. pp. 353-378.

Kirk, Geoffrey (ed.) 1982. *Schumacher on Energy*. London: Jonathan Cape.

Kumar, Krishan 1987. *Utopia and Anti-Utopia in Modern Times*. Oxford: Blackwell.

Lamb, Robert 1996. *Promising The Earth*. London: Routledge.

Lowe, Philip and Jane Goyder 1983. *Environment Groups In Politics*. London: Allen & Unwin.

Luckin, Bill 1990. *Questions of Power: Electricity and Environment in Inter-war Britain*. Manchester: Manchester University Press.

Marwick, Arthur 1998. *The Sixties*. Oxford: Oxford University Press.

Mazur, Alan 1981. *The Dynamics of Technical Controversy*. Washington DC: Communications Press.

McKay, George 1996. *Senseless Acts of Beauty: Cultures of resistance since the sixties*. London: Verso.

Morgan, Chris 1980. *The Shape of Futures Past: The story of prediction*. Exeter: Webb & Bower.

Nelson, Elizabeth 1989. *The British CounterCulture 1966-73: A study of the underground press*. London: Macmillan.

Nye, David 1994. *American Technological Sublime*. Cambridge, Ma: MIT Press.

Parrinder, Patrick (ed.) 1979. *Science Fiction: A critical guide*. London: Longman.

Patterson, Walt 1976. *Nuclear Power*. Harmondsworth: Penguin.

Patterson, Walt 1979. '"Environmental" Involvement in British Civil Nuclear Policy', *Journal of the Institute of Nuclear Engineering* 20/5:158-162.

Patterson, Walt 1984. 'A Decade of Friendship: The first ten years'. In *The Environmental Crisis*, ed. Des Wilson. London: Heinemann.

Patterson, Walt 1985. *Going Critical: An unofficial history of British nuclear power*. London: Paladin.

Pocock, R . F. 1977. *Nuclear Power*. Surrey: Gresham Press.

Roberts, Fred 1999. *60 Years of Nuclear History*. Charlbury. Oxon: Jon Carpenter.

Robinson, Kim Stanley (ed.) 1994. *Future Primitive: The new ecotopias*. New York: Tom Doherty.

Rome, Adam 2001. *The Bulldozer in the Countryside: Suburban sprawl and the rise of American environmentalism*. Cambridge: Cambridge University Press.

Rybczynski, Witold 1980. *Paper Heroes: A review of appropriate technology*. Garden City. New York: Anchor Press.

Scheer, Hermann 2002. *The Solar Economy: Renewable energy for a sustainable global future*. London: Earthscan.

Smil, Vaclav 2003. *Energy at the Crossroads: Global perspectives and uncertainties*. MIT Press.

Spiers, John 1974. *The Underground and Alternative Press in Britain*. Harvester Press.

Veldman, Meredith 1994. *Fantasy, the Bomb and the Greening of Britain: Romantic protest 1945-1980*. Cambridge: Cambridge University Press.

Walker, William 1999. *Nuclear Entrapment: THORP and the politics of commitment*. London: IPPR.

Wall, Derek 1999. *Earth First! and the Anti-Roads Movement: Radical environmentalism and comparative social movements*. London: Routledge.

Weart, Spencer 1988. *Nuclear Fear: A history of images*. London: Harvard University Press.

Welsh, Ian 2000. *Mobilizing Modernity: The nuclear moment*. London: Routledge.

Weston, Joe 1989. *The F.O.E. Experience*. Oxford Polytechnic, June 1989.

Williams, Roger 1980. *The Nuclear Power Decisions: British policies 1953-1978*. London: Croom Helm.

Winner, Langdon 1986. *The Whale and the Reactor: A search for limits in an age of high technology*. London: University of Chicago Press.

Wood, Barbara 1984. *Alias Papa: A life of Fritz Schumacher*. London: Jonathan Cape.

Wynne, Brian 1982. *Rationality and Ritual: the Windscale Inquiry and nuclear decisions in Britain*. London: British Society for History of Science.

Acronyms and Abbreviations

AEA	Atomic Energy Authority (UK)
AEC	Atomic Energy Commission (USA)
AGR	Advanced Gas-Cooled Reactor
AT	Alternative (or Appropriate) Technology
BNFL	British Nuclear Fuels Ltd. (UK)
CANTO	Concerns Against Nuclear Technology Organization
CEGB	Central Electricity Generating Board
CN	*Conservation News*
ConSoc	Conservation Society
FoE	Friends of the Earth (UK)
KS	*The Kidderminster Shuttle* (UK)
KT&SN	*Kidderminster Times and Stourport News* (UK)
NIMBY	'Not In My Back Yard!'
NIN	Nuclear Information Network (UK)
NNC	Network for Nuclear Concern
PERG	Political Ecology Research Group (UK)
PWR	Pressurized Water Reactor
RSPB	Royal Society for the Protection of Birds
RSPCA	Royal Society for the Prevention of Cruelty to Animals
SCRAM	Scottish Campaign to Resist the Atomic Menace
SEI	Society for Environmental Improvement
SERA	Socialist Environment and Resources Association (UK)
SGHWR	Steam Generating Heavy Water Reactor
TCPA	Town and Country Planning Association (UK)
THORP	Thermal Oxide Reprocessing Plant
TVA	Tennessee Valley Authority (USA)
UCS	Union of Concerned Scientists (USA)
WA	Windscale Appeal
WHO	World Health Organization
WCC	Worcestershire County Council

Footnotes

1   *Energy Utopianism and the rise of the anti-nuclear power movement in the UK*. PhD thesis. The Open University 2003.

2   Quotations of Alvin Weinberg from Nye 1994: 234 and Glenn Seaborg from Calder 1962: 61.

3   Thomas Sturgeon was later to write deeply pessimistic SF novels, *Memorial* (1946) and *Thunder and Roses* (1947), on the dangers and perhaps inevitability of nuclear holocaust.

4   Other anthologies were by the well-known SF writers Thomas Disch, *The Ruins of the Earth* (1971) and Roger Elwood & Virginia Kidd, *The Wounded Planet* (1973).

5   Ritchie Calder was a journalist, author and academic, later to become Lord Peter Ritchie-Calder. In the 1950s he was science editor of the London *News Chronicle*, then became in the 1960s Professor of International Relations at Edinburgh University. His fame and interest in food and population issues led him to be chosen as the president of the Conservation Society, where he gave the 1968 Presidential Address entitled 'Hell on Earth'.

6   The significance of milk and radiation can also be seen in the impact of the Windscale fire in October 1957. This received minimal adverse publicity until slightly radioactive milk was collected and dumped in the sea. One of the members of the WHO study group, according to Calder (1962) attributed this media concern to 'breast fed editors'.

7   Brower left the Sierra Club in 1969 to set up FoE, but after bitter internal battles with his staff he was forced to leave in 1984, and established the Earth Island Institute.

8   One of the most famous Renaissance utopias was *The City of the Sun* by Tommaso Capanella, published in 1623. For a summary see Carey 1999.

9   Geothermal power is often included amongst renewable sources of energy, though there is debate as to whether it is 'renewable' utilizing as it does heat from the interior of the earth. The fast breeder reactor and fusion energy are also claimed to be 'sustainable' energy as their fuel supply is practically unlimited but all nuclear power has been rejected by solar advocates on environmental and ethical grounds.

10 For a quotation from Fox on tidal, wave, wind, hydro and geothermal power, and use of solar collectors, see Clarke 1979.

11 Steve Baer was well-known as a dome builder at Drop City and the author of *Domes Cookbook* (1968).

12 In Britain the main work, in the style of *The Whole Earth Catalog*, was *Radical Technology* (1976), edited by Godfrey Boyle and Peter Harper – two participants in the British AT magazine, *Undercurrents*.

13 For a recent review of SF and ecological literature from the 1970s to 1990s see

Chapters 8 and 9 in *From Apocalypse to Way of Life* (2003) by Frederic Buell.

14   For a history of the use of SF as a means of political satire and criticism, see Chapter 4 in Morgan 1980. Notable writers who have used this technique are Aldous Huxley (1932), Ayn Rand (1938), Kurt Vonnegut (1952) and Michael Young (1958).

15   Hertzka was a well-known Viennese economist, who believed that the most desirable system for industrial production was the workers' co-operative, it being a very direct and participative form of capitalism. In order to promote his beliefs he wrote *Freeland: A Social Anticipation* (1891) which inspired many people to set up Freeland societies to put his ideas into practice (for further details see Morgan 1980).

16   Stephen Weart in his book *Nuclear Fear* mentions several instances of activists and scientists being deeply influenced by their reading of SF, including Helen Caldicott by Nevil Shute's *On The Beach* (1957), and Leo Szilard and Alvin Weinberg by H.G. Wells' *The World Set Free*.

17   Still going: http://www.liv.ac.uk/Philosophy/sf.html.

18   See http://www.euro.net/mark-space/bkLastAndFirstMen.html and a biography by Sam Moskowitz at:
http://www.geocities.com/Athens/Agora/7628/stapledon/bio.html.

19   J.D. Beresford was a prolific SF writer using a wide range of political and social themes. For details see Morgan 1980.

20   The ecological ideas of Richard Jefferies and W. H. Hudson are examined by Anna Bramwell (1989) in Chapter 7 on 'The Literary Ecologist'.

21   W. H. Hudson is described by Max Nicholson in *The New Environmental Age* (1987) as 'one of the most inspiring and influential pioneers of conservation and an early leader of the Royal Society for the Protection of Birds'.

22   See Introduction by John Fowles in the reissued edition of *After London, or Wild England* (1980).

23   The impact of the writing of C. S. Lewis and J. R. R. Tolkien on modern environmentalism is extensively examined in Section 1 of Meredith Veldman (1994).

24   The trilogy was *Out of the Silent Planet* (1938), *Perlanadra* (1943) and *That Hideous Strength* (1945). For discussion of these works see Veldman (1994) pp. 62-69, and for accounts of press comments see pp.102-104.

25   For the influences on Le Guin in 1968 see her introduction in the 1977 edition of *The Word for World is Forest*. For her use of fantasy and magic see her *Earthsea* trilogy 1969-73.

26   Much of the novel may be autobiographical as A. G. Street (1892-1966) was born in Wiltshire to a farming family, and after seven years in Canada, took over the farm in 1918.

27   Viscount Lymington (later known as the Earl of Portsmouth) was active in the organic movement and was writing on agricultural problems and solutions since 1932. His book *Famine in England* (1938) came out at the same time as Street's novel. For further details see numerous references by Anna Bramwell (1989) and by Philip Conford (2001).

28   The title of Stewart's novel comes from Ecclesiastes 1: 4: 'Men go and come, but

earth abides'. Interestingly an earlier novel with strong ecological concerns, *Already Walks Tomorrow* by A. G. Street, had its hero declare 'men come and go, but the land remains'.

29   Ironically Lord Robens, before being appointed chairman of the NCB in 1960, played a leading part in the construction of the Trawsfynydd nuclear power station as industrial relations adviser to the nuclear consortium, Atomic Power Constructions Ltd., who built it.

30   By coincidence protest at Dungeness power station was to feature as the first anti-nuclear event of two future groups, FoE in 1975 and Earth First! in 1991.

31   Over 40 years later in 2005 this area is again a place of conflict between conservationists and power developers – this time over plans to build the South-East's first onshore wind farm in Romney Marsh. Again the conflict pits government energy policy (and its green supporters) against wildlife and landscape protection and a public inquiry is to be held.

32   Written in February 1955 during his visit to Burma and published in 1962 by the Gandhian Institute of Studies, Varanasi, India in *Roots of Economic Growth*. Schumacher later used it in his first book, *Small is Beautiful*, and a companion paper 'Non-Violent Economics' was first published in *The Observer* on 21 August 1960.

33   The Christian religion was heavily influential in the development of the post-war organic tradition – see Chapter 11 'The Christian Context of Organic Husbandry' in Philip Conford's *The Origins of the Organic Movement*.

34   David Cooper was a radical 'anti-psychiatrist' and a well-known figure in the underground. He was a colleague of R. D. Laing, with whom he set up the Philadelphia Foundation, which in turn founded Kingsley Hall, a therapeutic community.

35   Articles about the London School of Non-Violence appeared in *It* no. 52 (March 1969), and Gandalf's Garden no. 6 (November 1969). For a statement of his views then, see his 'Non-Violence or Non-existence' published by Christian Action Publications 1969, and reviewed in *Resurgence* vol. 2, no. 6. Also note the cover of this issue (March/April 1969) which is in the form of a Captain Marvel cartoon with text dialogue on the nature of political consciousness – an interesting marriage of underground art with new left political thought. In late 1973 Satish Kumar became editor of Resurgence, a post he still holds.

36   Barry Commoner was the main speaker at the July 1969 Soil Association conference. He was an early supporter in the 1950s (according to Conford) of Jerome Rodale, the US organic farming researcher, who wrote on the dangers of chemical fertilizers and pesticides in 1949, and whose work was well-known to the Soil Association of which he became a vice-president.

37   Obviously a pseudonym, the Latin phrase 'pro bono publico' meaning for the good of the public.

38   Perhaps these are references to the controversy caused by the claims of Sternglass, which appeared on the front page of *The Times* on 27 January and on the same day in a Thames TV documentary 'Report on Pollution', in which he said: 'It now looks as if in countries like United States and England as many as one in two children that died before reaching age one died because of nuclear testing'.

His views were disputed the following day by Professor Joseph Rotblat in a Thames TV debate, and in *The Times* where Rotblat dismissed Sternglass' theories. The only support for Sternglass came from Arthur Tamplin who believed that 'the Sternglass work should cause us to take a more serious look at the potential effects of radiation from whatever source', and Tamplin opposed 'those who profess that exposure to low doses of radiation is harmless'. The book *Perils of the Peaceful Atom* by Curtis and Hogan had also been reviewed in *The Times* in March. The reviewer said that the book was 'clearly designed to do for nuclear pollution what Rachel Carson's book *Silent Spring* did for chemical pollution'. Most attention in the review was given to graphic passages from the book on nuclear waste described as 'a violent lethal mixture of short and long-lived isotopes' kept in storage tanks that 'could fail with catastrophic consequences'.

39   Both the accidents at Three Mile Island on 28 March 1979 and at Chernobyl on 26 April 1986 occurred in the early hours of the morning and resulted from human error.

40   Protest against the proposed twin AGR at Portskewett continued throughout the 1970s, becoming in the late 1970s part of the Severnside Alliance, a coalition of 25 local groups. The Portskewett and Sudbrook Action Group, nearest the proposed reactor, had in 1979 just 15 members, of which 8 were active.

41   The word 'hip' is said by Emmett Grogan, leader of the San Francisco Diggers, to stand for Haight Independent Proprietors, a group of Haight-Ashbury shopkeepers profiting from the thousands of young people flocking to San Francisco in 1967, encouraged by media reports of a 'Summer of Love'. Grogan in his autobiography *Ringolevio: A Life Played for Keeps* (1972), passionately denounced the media and 'the way it catered to a hip, moneyed class by refusing to reveal the overall grime of the Haight-Ashbury reality... absolute bullshit implicit in the psychedelic transcendentalism promoted by the self-proclaimed, media-fabricated shamans who espoused the tune-in, drop-out, jerk-off ideology of Leary and Alpert'. See a US website devoted to the history of hippies http://wild-bohemian.com/#hippies.

42   This is a paraphrase of his views. In his book *Playpower* Neville actually wrote: 'Tomorrow you may be paid NOT to work – can you take it?' (p219); Like a child learning its first steps, members of the Underground are learning how to live in that future world where work is rendered obsolete' (p 222) and 'Technology will provide basic needs' (p 221).

43   For a list and comments on UK publication see *Directory of Alternative Periodicals* (1st edition 1970, 2nd 1972, 3rd 1974) compiled by John Noyce, a noted alternative press archivist. Later he compiled *Directory of British Alternative Periodicals 1965-1974* (1979), a more complete listing. There was also *Alternative Press Digest* (no. 1 June, 1974; no. 3 May 1975), a selection of articles reprinted from alternative press, published by Magic Ink, Margate by Ian King, an ex-BIT worker. He also published the *UAPS (the Underground / Alternative Press Service) Catalogue of Alternative Papers*, and *Magic Ink*, an internal newsletter of UAPS/Europe sent to papers across the world. In the USA there was the *Alternative Press Index* – an index to alternative and underground publications

from Alternative Press Centre, Baltimore, with an index of articles by subject. For an index of the contents of UK environmental magazines in the 1970s see my *Guide to Environmental Resources* (1975) and *Alternative Technology Directory* (1978).

44   Another big influence on the mystic scene was John Michell, author of the classic *The View over Atlantis* (1973) and writer for IT on ley lines.

45   Many of these articles on crafts were reproduced in the book by its editors, Andy Pittaway and Bernard Scofield, *Country Bizarre's Country Bazaar* (1974).

46   Joy Farren was wife of Mike Farren, the editor of *IT* who according to Jonathan Green's *Days in the Life* (1988) was 'one of the movement's most out-spoken figures'. While Mike Farren receives copious mention in any history of the underground, Joy receives none.

47   For a personal account of *Snail* by Peter Blake, see Clem Gorman's *People Together* (1975).

48   From June 1973 it was renamed *Whole Earth News*, and appeared intermittently published by Tony Brantingham, then living in Shropshire. It is not to be confused with the magazine *Whole Earth*, published in Brighton from 1974 to 1979 or the new left paper *Black Dwarf*.

49   Dave Cunliffe was a poet, vegan and anarchist and editor of the poetry magazine Poetmeat. He had attracted widespread publicity and literary support after being fined in 1965 for publishing *The Golden Convolvulus*, which was judged 'indecent' but not 'obscene'. He was an early (and regular) contributor to *Resurgence* (see *Resurgence* vol. 2 no. 12 for biographical details).

50   According to Nicholas Albery there were only about 90 people in the six communes affiliated to the Communes Movement in 1970 (Progress report from the Commune Movement, *Resurgence* vol. 3, no. 2 (July/August 1970).

51   Sid Rawles was in the mid-1960s the leader of the Communist Party in Slough and NUPE shop steward. He then became very active in the 'free festivals' movement in the 1970s and early 1980s, and joined the Ecology Party in 1980.

52   The word 'Survival' was very popular and a much abused term since the mid-1960s, featuring in book titles, such *as Science and Survival* by Barry Commoner (1966), The *Survival Scrapbooks* (1972), and *The Survivalists* by Patrick Rivers (1971); in eco-press titles, *Towards Survival*; in group names: Action for Survival, and Movement for Survival (both founded in 1972); in TV programmes, like *Question of Survival* (14 December 1971); and in articles in the alternative press, as with *Survival?* by Joy Farren (*Ink* November 1971),

53   Rona left in early 1974 and was replaced by Ellen Kistler, and the commune shrank to three people, Dick, Ellen and Paul Richards.

54   The 'environmental justice' movement that originated in the USA in the late 1980s argues that living in a poor quality environment (often next to hazardous waste dumps) is not a matter of choice but of power interests.

55   For a later account of women activists opposing nuclear power in California in the late 1970s and early 1980s, see John Wills' article 'Talking Atoms: Anti-Nuclear Protest at Diablo Canyon, California 1977-1984' in *Oral History* Autumn 2000.

56   Perhaps the first book written by a woman activist was *Ground for Concern:*

Australia's Uranium and Human Survival by Mary Elliott, a longstanding Australian anti-nuclear activist, in 1977.

57   This challenge was to culminate in the mass trespass on Kinder Scout in 1932 but not achieve victory until the implementation of the Right to Roam Act in 2004.

58   Very little has been written about the Dwarfs, though they had an extensive network of branches, their own magazine *Dwarf News* and were active in 1971 in protesting about air pollution (see *Ink* no. 12 and *Frendz* no. 13). They adopted a form of 'anarchist ecology', modeled on previous groups such as the Dutch Provos, and were well-known to hippies for their festivals and concerts. However a contemporary and similar group, the Diggers, is better known though it received less coverage in the alternative press, mainly because of the antics of Sid Rawles, its self-appointed leader.

59   One specific reason given for David Brower's resignation from the Sierra Club was the board's refusal to oppose the Diablo Canyon nuclear power station. Brower had initially supported nuclear power as a 'smart alternative' to hydropower dams but by the late 1960s was opposing it on grounds of safety and economics.

60   Richard Sandbrook was former student union president at the University of East Anglia, and an environmental activist within the NUS.

61   It was also at Dungeness nuclear power station where Earth First! held their first public action in the UK in 1991, when about 60 people from a variety of peace and anti-nuclear groups such as the Brighton Peace Centre, Sea Action, and Dungeness Action Society of Hastings blockaded the station. According to *Green Anarchist* in 1991 the Dungeness area contained 'the best example of a cuspate foreland in the world… home to over 600 species of flora and fauna, some of which are rare'. An exactly similar point had been made by protesters at the public inquiry in 1958. Protest it seems is rarely novel, arguments and methods are handed down from one generation of activists to another. The results of protest, generally failure, are unfortunately depressingly similar.

62   This tactic of getting usually pro-nuclear newspapers to run critical stories on the bureaucratic shortcomings of nuclear power was copied by Walt Patterson and Amory Lovins of FoE in Britain in 1973-4. At a time of large media coverage of the 'energy crisis' journalists were hungry for energy stories, particularly of the 'cover-up' type. Walt Patterson – a vocal critic of the AEA – was able to exploit this desire, such as when he leaked the story that the AEA, perhaps out of defensiveness, requested him not to ask questions at their press conference.

63   This observation is not made cynically but reflects the fact that NGOs rely for their survival on membership fees and donations from the public and thus have to continually find issues that appeal to their constituencies.

64   Patterson made similar comments in 1973, when he wrote in his Preface to his first book *Nuclear Reactors* that 'a nuclear reactor is scientifically no more mysterious than a transistor radio' and of the need to dispel the impression that nuclear reactors can only be understood and discussed by the 'experts'.

65   Davy was a long-time writer on this theme, having published in 1962 an article entitled 'The deadly wastes of nuclear power'.

66   The Royal Commission on Environmental Pollution inquiry on *Nuclear Power and the Environment* was announced on 19 March1974, and published its 6th report on 22 September 1976; often referred to as *The Flowers Report*.

67   John Fremlin, professor of Applied Radioactivity at Birmingham University was a ConSoc member and a prolific letter writer in support of nuclear power. He was also, according to Ian Breach, a long-time CND supporter and helped found BSSRS. In 1976 he was appointed a consultant on radiation risk to the Cumbria County Council, and gave evidence to the Windscale Inquiry. He later wrote *Power Production: what are the risks?* (1987).

68   The PWR (pressurized water reactor) is a type of LWR (light water reactor), the other being the BWR (boiling water reactor); both types were developed in the USA. The SGHWR (steam generating heavy water reactor) was a British design with only one prototype built at Winfrith, Dorset.

69   John Price was a researcher for FoE on energy analysis of nuclear power – see John Price and Amory Lovins' *Dynamic Energy Analysis and Nuclear Power* (1974). He left FoE in the 1970s and went to Australia. Interestingly neither Lamb nor Patterson mention his work for FoE.

70   His wife, Mrs. G. J. Acland, was one of the only two county councillors (out of 98) who were sympathetic to the Windscale opponents.

71   Brian Wynne was a lecturer in the School of Independent Studies at Lancaster University. After the Windscale Inquiry, he wrote a series of articles for *New Scientist* and *Rationality and Ritual* (1982) on nuclear decision-making and the role of public inquiries. It contains, in Chapter 6, much detail about the conflicts between the groups, represented at the inquiry. Wynne has continued to write on Windscale and radiation issues.

72   Sir Frank Layfield was to achieve widespread publicity in the 1980s as the inspector at the Sizewell Inquiry.

73   Ten years later, the proposed Hinkley C PWR galvanized local opposition, which became the Stop Hinkley Expansion (SHE) group.

74   According to Ian Breach, Peter Taylor was a former social anthropologist who had spent six years in industry and teaching. His nuclear expertise had come from acting as an advisor to local citizens' groups opposed to the construction of a reprocessing plant at Aschendorf, West Germany and from working alongside university lobbyists from Bremen.

75   John Tyme was a well-known, even notorious, anti-roads campaigner who achieved much media publicity through his direct action activities, including disrupting dozens of road inquiries in Britain. His campaigning resulted in personal tragedy – he suffered a nervous breakdown and retreated to a monastery. According to his speech at the 1986 ConSoc national conference his fifteen years of impassioned efforts 'have mostly resulted in complete failure, much to his anguish'.

76   Transport 2000 was formed in 1973 at the initiative of the National Union of Railwaymen (NUR) as a federation of groups, including environmental ones, in favour of developing the railways. The NUR provided the secretariat and much of the finance for Transport 2000.

77   Ironically over 25 years later, in October 2004, Hugh Montefiore had to resign

from the board of FoE after supporting nuclear power as the only viable option to deal with global warming.

78 Sir Kelvin Spencer was a former chief scientist with the Ministry of Power in the 1950s. According to Walt Patterson he had become a 'fiercely outspoken opponent of official British nuclear policy' after having been an 'enthusiastic supporter of nuclear power'. He gave evidence for the SEI at the Windscale Inquiry.

79 Charles Wakstein had published an article critical of Windscale's safety procedures in the *Whitehaven News* in December 1975, which was enthusiastically reprinted in *Undercurrents* no. 16, with a further supportive article for his views in no. 18 by Martyn Partridge.

80 See Walker 1999 pp. 93-94 for discussion of this option.

81 For the latest news on the nuclear industry, visit: http://www.guardian.co.uk/nuclear/0,2759,181325,00.html.

82 For a statement of FoE's position on direct action, see letter by Tom Burke to *Vole* in October 1978, after *Vole*'s editorial in no. 12 highlighted FoE's 'conspicuous lack of enthusiasm' for direct action.

83 The most famous one being the 1962 utopian novel *Island* by Aldous Huxley.

84 This opposition by the CPRW is long-standing. At the Connah's Quay inquiry in 1971 the CPRW supported the construction of nuclear power stations in industrial areas, rather than in 'unspoilt' rural coastal regions.

Index